Books are to be returned on or before
last date below.

PRINCIPLES AND APPLICATIONS IN ENGINEERING SERIES

Biotechnology for Biomedical Engineers

PRINCIPLES AND APPLICATIONS IN ENGINEERING SERIES

Biotechnology for Biomedical Engineers

Edited by

MARTIN L. YARMUSH
MEHMET TONER
ROBERT PLONSEY
JOSEPH D. BRONZINO

CRC PRESS

Boca Raton   London   New York   Washington, D.C.

Library of Congress Cataloging-in-Publication Data

Biotechnology for biomedical engineers / Martin L. Yarmush ... [et al.].
    p. cm. (Principles and applications in engineering)
    ISBN 0-8493-1811-4 (alk. paper)
    1. Biotechnology I. Yarmush, Martin, L. II. Series.

TP248.2.B5517 2003
660.6—dc21   2002041504

Visit the CRC Press Web site at www.crcpress.com

© 2003 by CRC Press LLC

No claim to original U.S. Government works
International Standard Book Number ISBN 0-8493-1811-4
Library of Congress Card Number 2002041504
Printed in the United States of America 1 2 3 4 5 6 7 8 9 0
Printed on acid-free paper

Preface

The human genome project has altered the very nature of research and development related to the treatment of disease and, in the process, has revolutionized the field of "biotechnology." Pioneering work in genomics, for example, has led to the development of sophisticated techniques for determining differential gene-expression patterns (transcriptomics) resulting from genetic makeup, disease slate or influence from external factors. This book — *Biotechnology for Biomedical Engineers* — takes the sections most relevant to this important topic from the second edition of the *Biomedical Engineering Handbook* published in 2000. Since it is important for individuals engaged in this field to understand the fundamentals of physiology, this handbook opens with a section on Physiologic Systems, edited by Robert Plonsey, which provides an overview of the major physiologic systems of current interest to biomedical engineers, namely the cardiovascular, endocrine, nervous, visual, auditory, gastrointestinal and respiratory systems. It is important to note that this section is written at an introductory and tutorial level. However, since this book has been prepared for the biomedical engineering community, mathematical descriptions are not avoided.

In the subsequent chapters, the major editors, Drs. Martin Yarmush and Mehmet Toner, have assembled material that covers most topics in biotechnology that might interest the practicing biomedical engineer. During the past two decades, the field of biotechnology in the advent of recombinant DNA technology, monoclonal antibody technology, and new technologies for studying and handling cells and tissues, has gone through a tremendous resurgence in a wide range of applications pertinent to industry, medicine, and science in general. Some of these new ideas, concepts, and technologies are covered in this handbook.

With this in mind, the *Biotechnology for Biomedical Engineers Handbook* presents:

- Approaches and techniques to manipulate genetic materials. This capability, which provides the practitioner with the potential to generate new proteins with improved biochemical and physio-chemical properties, has led to the formation of the field of **protein engineering**.

- The field of monoclonal antibody production in terms of its basic technology, diverse applications, and ways that the field of recombinant DNA technology is currently "reshaping" some of the earlier constructs.

- Applications of nucleic acid chemistry, as well as the burgeoning field of antisense technology, with emphasis on basic techniques and potential applications to AIDS and cancer.

- The computational, chemical, and machine tools that are being developed and refined for genome analysis.

- The fundamentals of applied virology in which viral vaccines and viral-mediated gene therapy are the main foci.

- Important aspects of cell structure and function, emphasizing a common approach toward quantitative analysis of cell behavior in order to develop the principles for cell growth and function.

By viewing the world of biotechnology through the use of proteins and nucleic acids and through viruses to cells, today's biomedical engineer will hopefully be prepared to meet the challenge of participating in the greater field of biotechnology.

Joseph D. Bronzino

Advisory Board

Contributors

John G. Aunins
Merck Research Laboratories
Rahway, New Jersey

Berj L. Bardakjian
University of Toronto
Toronto, Canada

Ewart R. Carson
City University
London, United Kingdom

Ben M. Clopton
University of Washington
Seattle, Washington

Derek G. Cramp
City University
London, United Kingdom

Leslie A. Geddes
Purdue University
West Lafayette, Indiana

Arthur T. Johnson
University of Maryland
College Park, Maryland

Robert Kaiser
University of Washington
Seattle, Washington

Douglas A. Lauffenburger
Massachusetts Institute of
    Technology
Cambridge, Massachusetts

Christopher G. Lausted
University of Maryland
College Park, Maryland

Joseph M. Le Doux
Center for Engineering in Medicine,
    and Surgical Services,
    Massachusetts General Hospital,
    Harvard Medical School, and the
    Shriners Burns Hospital
Cambridge, Massachusetts

Ann L. Lee
Merck Research Laboratories
Rahway, New Jersey

**Evangelia Micheli-
Tzanakou**
Rutgers University
Piscataway, New Jersey

Jeffrey R. Morgan
Center for Engineering in Medicine,
    and Surgical Services,
    Massachusetts General Hospital,
    Harvard Medical School, and the
    Shriners Burns Hospital
Cambridge, Massachusetts

Charles M. Roth
Center for Engineering in Medicine,
    Massachusetts General Hospital,
    Harvard Medical School, and the
    Shriners Burns Hospital
Cambridge, Massachusetts

Alan J. Russell
University of Pittsburgh
Pittsburgh, Pennsylvania

John Schenck
General Electric Corporate Research
    and Development Center
Schenectady, New York

Francis A. Spelman
University of Washington
Seattle, Washington

George Stetten
Duke University
Durham, North Carolina

Srikanth Sundaram
Rutgers University
Piscataway, New Jersey

Chenzhao Vierheller
University of Pittsburgh
Pittsburgh, Pennsylvania

David B. Volkin
Merck Research Laboratories
Rahway, New Jersey

S. Patrick Walton
Center for Engineering in Medicine,
    Massachusetts General Hospital,
    Harvard Medical School, and the
    Shriners Burns Hospital
Cambridge, Massachusetts

Martin L. Yarmush
Massachusetts General Hospital,
    Harvard Medical School, and the
    Shriners Burns Hospital
Cambridge, Massachusetts

Craig Zupke
Massachusetts General Hospital and
    the Shriners Burns Institute
Cambridge, Massachusetts

Contents

I

Physiologic Systems

Robert Plonsey
Duke University

T HE CONTENT OF THIS HANDBOOK is devoted to the subject of *biomedical engineering*. We understand biomedical engineering to involve the application of engineering science and technology to problems arising in medicine and biology. In principle, the intersection of each engineering discipline (i.e., electrical, mechanical, chemical, etc.) with each discipline in medicine (i.e., cardiology, pathology, neurology, etc.) or biology (i.e., biochemistry, pharmacology, molecular biology, cell biology, etc.) is a potential area of biomedical engineering application. As such, the discipline of biomedical engineering is potentially very extensive. However, at least to date, only a few of the aforementioned "intersections" contain active areas of research and/or development. The most significant of these are described in this handbook.

While the application of engineering expertise to the life sciences requires an obvious knowledge of contemporary technical theory and its applications, it also demands an adequate knowledge and understanding of relevant medicine and biology. It has been argued that the most challenging part of finding engineering solutions to problems lies in the formulation of the solution in engineering terms. In biomedical engineering, this usually demands a full understanding of the life science substrates as well as the quantitative methodologies.

This section is devoted to an overview of the major physiologic systems of current interest to biomedical engineers, on which their work is based. The overview may contain useful definitions, tables of basic physiologic data, and an introduction to the literature. Obviously these chapters must be extremely brief. However, our goal is an introduction that may enable the reader to clarify some item of interest or to indicate a way to pursue further information. Possibly the reader will find the greatest value in the references to more extensive literature.

This section contains seven chapters, and these describe each of the major organ systems of the human body. Thus we have chapters describing the cardiovascular, endocrine, nervous, visual, auditory, gastrointestinal, and respiratory systems. While each author is writing at an introductory and tutorial level, the audience is assumed to have some technical expertise, and consequently, mathematical descriptions are not avoided. All authors are recognized as experts on the system that they describe, but all are also biomedical engineers.

The authors in this section noted that they would have liked more space but recognized that the main focus of this handbook is on "engineering." Hopefully, readers will find this introductory section helpful to their understanding of later chapters and, as noted above, to at least provide a starting point for further investigation into the life sciences.

1

An Outline of Cardiovascular Structure and Function

Daniel J. Schneck
Virginia Polytechnic Institute and State University

Because not every cell in the human body is near enough to the environment to easily exchange with it mass (including nutrients, oxygen, carbon dioxide, and the waste products of metabolism), energy (including heat), and momentum, the physiologic system is endowed with a major highway network — organized to make available thousands of miles of access tubing for the transport to and from a different neighborhood (on the order of 10 μm or less) of any given cell whatever it needs to sustain life. This highway network, called the *cardiovascular system,* includes a pumping station, the heart; a working fluid, blood; a complex branching configuration of distributing and collecting pipes and channels, blood vessels; and a sophisticated means for both intrinsic (inherent) and extrinsic (autonomic and endocrine) control.

1.1   The Working Fluid: Blood

Accounting for about 8 ± 1% of total body weight, averaging 5200 ml, blood is a complex, heterogeneous suspension of formed elements — the *blood cells,* or *hematocytes* — suspended in a continuous, straw-colored fluid called *plasma.* Nominally, the composite fluid has a mass density of 1.057 ± 0.007 g/cm³, and it is three to six times as viscous as water. The hematocytes (Table 1.1) include three basic types of cells: red blood cells (erythrocytes, totaling nearly 95% of the formed elements), white blood cells (leukocytes, averaging <0.15% of all hematocytes), and platelets (thrombocytes, on the order of 5% of all blood cells). Hematocytes are all derived in the active ("red") bone marrow (about 1500 g) of adults from undifferentiated stem cells called *hemocytoblasts,* and all reach ultimate maturity via a process called *hematocytopoiesis.*

The primary function of erythrocytes is to aid in the transport of blood gases — about 30 to 34% (by weight) of each cell consisting of the oxygen- and carbon dioxide–carrying protein hemoglobin (64,000 ≤ MW ≤68,000) and a small portion of the cell containing the enzyme carbonic anhydrase, which catalyzes the reversible formation of carbonic acid from carbon dioxide and water. The primary function of leukocytes is to endow the human body with the ability to identify and dispose of foreign substances such as infectious organisms) that do not belong there — agranulocytes (lymphocytes and monocytes)

TABLE 1.1 Hematocytes

Cell Type	Number Cells per mm³ Blood*	Corpuscular Diameter (µm)*	Corpuscular Surface Area (µm²)*	Corpuscular Volume (µm³)*	Mass Density (g/cm³)*	Percent Water*	Percent Protein*	Percent Extractives*†
Erythrocytes (red blood cells)	4.2–5.4 × 10⁶ ♀ 4.6–6.2 × 10⁶ ♂ (5 × 10⁶)	6–9 (7.5) Thickness 1.84–2.84 "Neck" 0.81–1.44	120–163 (140)	80–100 (90)	1.089–1.100 (1.098)	64–68 (66)	29–35 (32)	1.6–2.8 (2)
Leukocytes (white blood cells)	4000–11000 (7500)	6–10	300–625	160–450	1.055–1.085	52–60 (56)	30–36 (33)	4–18 (11)
Granulocytes								
Neutrophils: 55–70% WBC (65%)	2–6 × 10³ (4875)	8–8.6 (8.3)	422–511 (467)	268–333 (300)	1.075–1.085 (1.080)	—	—	—
Eosinophils: 1–4% WBC (3%)	45–480 (225)	8–9 (8.5)	422–560 (491)	268–382 (321)	1.075–1.085 (1.080)	—	—	—
Basophils: 0–1.5% WBC (1%)	0–113 (75)	7.7–8.5 (8.1)	391–500 (445)	239–321 (278)	1.075–1.085 (1.080)	—	—	—
Agranulocytes								
Lymphocytes: 20–35% WBC (25%)	1000–4800 (1875)	6.75–7.34 (7.06)	300–372 (336)	161–207 (184)	1.055–1.070 (1.063)	—	—	—
Monocytes: 3–8% WBC (6%)	100–800 (450)	9–9.5 (9.25)	534–624 (579)	382–449 (414)	1.055–1.070 (1.063)	—	—	—
Thrombocytes (platelets)	(1.4 ♂), 2.14 (♀)–5 ×10⁵ (2.675 × 10⁵)	2–4 (3) Thickness 0.9–1.3	16–35 (25)	5–10 (7.5)	1.04–1.06 (1.05)	60–68 (64)	32–40 (36)	Neg.

*Normal physiologic range, with "typical" value in parentheses.
†Extractives include mostly minerals (ash), carbohydrates, and fats (lipids).

essentially doing the "identifying" and granulocytes (neutrophils, basophils, and eosinophils) essentially doing the "disposing." The primary function of platelets is to participate in the blood-clotting process.

Removal of all hematocytes from blood centrifugation or other separating techniques leaves behind the aqueous (91% water by weight, 94.8% water by volume), saline (0.15 N) suspending medium called *plasma* — which has an average mass density of 1.035 ± 0.005 g/cm³ and a viscosity 1½ to 2 times that of water. Some 6.5 to 8% by weight of plasma consists of the plasma proteins, of which there are three major types — albumin, the globulins, and fibrinogen — and several lesser prominence (Table 1.2).

TABLE 1.2 Plasma

Constituent	Concentration Range (mg/dl plasma)	Typical Plasma Value (mg/dl)	Molecular Weight Range	Typical Value	Typical Size (nm)
Total protein, 7% by weight	6400–8300	7245	21,000–1,200,000	—	—
Albumin (56% TP)	2800–5600	4057	66,500–69,000	69,000	15 × 4
α_1-*Globulin* (5.5% TP)	300–600	400	21,000–435,000	60,000	5–12
α_2-*Globulin* (7.5% TP)	400–900	542	100,000–725,000	200,000	50–500
β-*Globulin* (13% TP)	500–1230	942	90,000–1,200,000	100,000	18–50
γ-*Globulin* (12% TP)	500–1800	869	150,000–196,000	150,000	23 × 4
Fibrinogen (4% TP)	150–470	290	330,000–450,000	390,000	(50–60) × (3–8)
Other (2% TP)	70–210	145	70,000–1,000,000	200,000	(15–25) × (2–6)
Inorganic ash, 0.95% by weight	930–1140	983	20–100	—	— (Radius)
Sodium	300–340	325	—	22.98977	0.102 (Na^+)
Potassium	13–21	17	—	39.09800	0.138 (K^+)
Calcium	8.4–11.0	10	—	40.08000	0.099 (Ca^{2+})
Magnesium	1.5–3.0	2	—	24,30500	0.072 (Mg^{2+})
Chloride	336–390	369	—	35.45300	0.181 (Cl^-)
Bicarbonate	110–240	175	—	61.01710	0.163 (HCO_3^-)
Phosphate	2.7–4.5	3.6	—	95.97926	0.210 (HPO_4^{2-})
Sulfate	0.5–1.5	1.0	—	96.05760	0.230 (SO_4^{2-})
Other	0–100	80.4	20–100	—	0.1–0.3
Lipids (fats), 0.80% by weight	541–1000	828	44,000–3,200,000	= Lipoproteins	Up to 200 or more
Cholesterol (34% TL)	12–105 "free" 72–259 esterified, 84–364 "total"	59 224 283	386.67	Contained mostly in intermediate to LDL β-lipoproteins; higher in women	
Phospholipid (35% TL)	150–331	292	690–1010	Contained mainly in HDL to VHDL α_1-lipoproteins	
Triglyceride (26% TL)	65–240	215	400–1370	Contained mainly in VLDL α_2-lipoproteins and chylomicrons	
Other (5% TL)	0–80	38	280–1500	Fat-soluble vitamins, prostaglandins, fatty acids	
Extractives, 0.25% by weight	200–500	259	—	—	—
Glucose	60–120, fasting	90	—	180.1572	0.86 D
Urea	20–30	25	—	60.0554	0.36 D
Carbohydrate	60–105	83	180.16–342.3	—	0.74–0.108 D
Other	11–111	61	—	—	—

The primary functions of albumin are to help maintain the osmotic (oncotic) transmural pressure differential that ensures proper mass exchange between blood and interstitial fluid at the capillary level and to serve as a transport carrier molecule for several hormones and other small biochemical constituents (such as some metal ions). The primary function of the globulin class of proteins is to act as transport carrier molecules (mostly of the α and β class) for large biochemical substances, such as fats (lipoproteins) and certain carbohydrates (muco- and glycoproteins) and heavy metals (mineraloproteins), and to work together with leukocytes in the body's immune system. The latter function is primarily the responsibility of the γ class of immunoglobulins, which have antibody activity. The primary function of fibrinogen is to work with thrombocytes in the formation of a blood clot — a process also aided by one of the most abundant of the lesser proteins, prothrombin (MW ≃ 62,000).

Of the remaining 2% or so (by weight) of plasma, just under half (0.95%, or 983 mg/dl plasma) consists of minerals (inorganic ash), trace elements, and electrolytes, mostly the cations sodium, potassium, calcium, and magnesium and the anions chlorine, bicarbonate, phosphate, and sulfate — the latter three helping as buffers to maintain the fluid at a slightly alkaline pH between 7.35 and 7.45 (average 7.4). What is left, about 1087 mg of material per deciliter of plasma, includes: (1) mainly (0.8% by weight) three major types of fat, i.e., cholesterol (in a free and esterified form), phospholipid (a major ingredient of cell membranes), and triglyceride, with lesser amounts of the fat-soluble vitamins (A, D, E, and K), free fatty acids, and other lipids, and (2) "extractives" (0.25% by weight), of which about two-thirds includes glucose and other forms of carbohydrate, the remainder consisting of the water-soluble vitamins (B-complex and C), certain enzymes, nonnitrogenous and nitrogenous waste products of metabolism (including urea, creatine, and creatinine), and many smaller amounts of other biochemical constituents — the list seeming virtually endless.

Removal from blood of all hematocytes and the protein fibrinogen (by allowing the fluid to completely clot before centrifuging) leaves behind a clear fluid called *serum*, which has a density of about 1.018 ± 0.003 g/cm³ and a viscosity up to 1½ times that of water. A glimpse of Tables 1.1 and 1.2, together with the very brief summary presented above, nevertheless gives the reader an immediate appreciation for why blood is often referred to as the "river of life." This river is made to flow through the vascular piping network by two central pumping stations arranged in series: the left and right sides of the human heart.

1.2   The Pumping Station: The Heart

Barely the size of the clenched fist of the individual in whom it resides — an inverted, conically shaped, hollow muscular organ measuring 12 to 13 cm from base (top) to apex (bottom) and 7 to 8 cm at its widest point and weighing just under 0.75 lb (about 0.474% of the individual's body weight, or some 325 g) — the human heart occupies a small region between the third and sixth ribs in the central portion of the thoracic cavity of the body. It rests on the diaphragm, between the lower part of the two lungs, its base-to-apex axis leaning mostly toward the left side of the body and slightly forward. The heart is divided by a tough muscular wall — the interatrial-interventricular septum — into a somewhat crescent-shaped right side and cylindrically shaped left side (Fig. 1.1), each being one self-contained pumping station, but the two being connected in series. The left side of the heart drives oxygen-rich blood through the aortic semilunar outlet valve into the *systemic circulation*, which carries the fluid to within a differential neighborhood of each cell in the body — from which it returns to the right side of the heart low in oxygen and rich in carbon dioxide. The right side of the heart then drives this oxygen-poor blood through the pulmonary semilunar (pulmonic) outlet valve into the *pulmonary circulation*, which carries the fluid to the lungs — where its oxygen supply is replenished and its carbon dioxide content is purged before it returns to the left side of the heart to begin the cycle all over again. Because of the anatomic proximity of the heart to the lungs, the right side of the heart does not have to work very hard to drive blood through the pulmonary circulation, so it functions as a low-pressure ($P \leq 40$ mmHg gauge) pump compared with the left side of the heart, which does most of its work at a high pressure (up to 140 mmHg gauge or more) to drive blood through the entire systemic circulation to the furthest extremes of the organism.

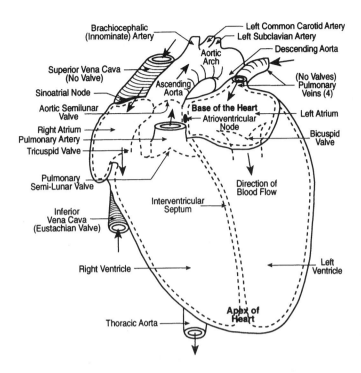

FIGURE 1.1.   Anterior view of the human heart showing the four chambers, the inlet and outlet valves, the inlet and outlet major blood vessels, the wall separating the right side from the left side, and the two cardiac pacing centers — the sinoatrial node and the atrioventricular node. Boldface arrows show the direction of flow through the heart chambers, the valves, and the major vessels.

Each cardiac (heart) pump is further divided into two chambers: a small upper receiving chamber, or atrium (auricle), separated by a one-way valve from a lower discharging chamber, or ventricle, which is about twice the size of its corresponding atrium. In order of size, the somewhat spherically shaped left atrium is the smallest chamber — holding about 45 ml of blood (at rest), operating at pressures on the order of 0 to 25 mmHg gauge, and having a wall thickness of about 3 mm. The pouch-shaped right atrium is next (63 ml of blood, 0 to 10 mmHg gauge of pressure, 2-mm wall thickness), followed by the conical/cylindrically shaped left ventricle (100 ml of blood, up to 140 mmHg gauge of pressure, variable wall thickness up to 12 mm) and the crescent-shaped right ventricle (about 130 ml of blood, up to 40 mmHg gauge of pressure, and a wall thickness on the order of one-third that of the left ventricle, up to about 4 mm). All together, then, the heart chambers collectively have a capacity of some 325 to 350 ml, or about 6.5% of the total blood volume in a "typical" individual — but these values are nominal, since the organ alternately fills and expands, contracts, and then empties as it generates a *cardiac output.*

During the 480-ms or so filling phase — diastole — of the average 750-ms cardiac cycle, the inlet valves of the two ventricles (3.8-cm-diameter tricuspid valve from right atrium to right ventricle; 3.1-cm-diameter bicuspid or mitral valve from left atrium to left ventricle) are open, and the outlet valves (2.4-cm-diameter pulmonary valve and 2.25-cm-diameter aortic semilunar valve, respectively) are closed — the heart ultimately expanding to its end-diastolic-volume (EDV), which is on the order of 140 ml of blood for the left ventricle. During the 270-ms emptying phase — systole — electrically induced vigorous contraction of cardiac muscle drives the intraventricular pressure up, forcing the one-way inlet valves closed and the unidirectional outlet valves open as the heart contracts to its end-systolic-volume (ESV), which is typically on the order of 70 ml of blood for the left ventricle. Thus the ventricles normally empty about half their contained volume with each heart beat, the remainder being termed the *cardiac reserve volume.* More generally, the difference between the *actual* EDV and the *actual* ESV, called the *stroke volume* (SV), is the volume of blood expelled from the heart during each systolic interval, and the ratio of SV to EDV is called the *cardiac ejection fraction,* or *ejection ratio* (0.5 to 0.75 is normal, 0.4 to 0.5 signifies mild cardiac damage, 0.25 to 0.40 implies moderate heart damage, and <0.25 warms of severe damage to the heart's pumping ability). If the stroke volume is multiplied by the number of systolic intervals per minute, or heart (HR), one obtains the total cardiac output (CO):

$$CO = HR \times \left(EDV - ESV\right). \qquad (1.1)$$

Dawson [1991] has suggested that the cardiac output (in milliliters per minute) is proportional to the weight W (in kilograms) of an individual according to the equation,

$$CO - 224W^{3/4}, \qquad (1.2)$$

and that "normal" heart rate obeys very closely the relation,

$$HR = 229W^{-1/4}. \qquad (1.3)$$

For a "typical" 68.7-kg individual (blood volume = 5200 ml), Equations (1.1), (1.2), and (1.3) yield CO = 5345 ml/min, HR = 80 beats/min (cardiac cycle period = 754 ms) and SV = CO/HR = $224W^{3/4}$/$229W^{-1/4}$ = $0.978W$ = 67.2 ml/beat, which are very reasonable values. Furthermore, assuming this individual lives about 75 years, his or her heart will have cycled over 3.1536 billion times, pumping a total of 0.2107 billion liters of blood (55.665 million gallons, or 8134 quarts per day) — all of it emptying into the circulatory pathways that constitute the vascular system.

1.3   The Piping Network: Blood Vessels

The vascular system is divided by a microscopic capillary network into an upstream, high-pressure, efferent arterial side (Table 1.3) — consisting of relatively thick-walled, viscoelastic tubes that carry blood away from the heart — and a downstream, low-pressure, afferent venous side (Table 1.4) — consisting of correspondingly thinner (but having a larger caliber) elastic conduits that return blood back to the heart. Except for their differences in thickness, the walls of the largest arteries and veins consist of the same three distinct, well-defined, and well-developed layers. From innermost to outermost, these layers are (1) the thinnest *tunica intima,* a continuous lining (the vascular endothelium) consisting of a single layer of simple squamous (thin, sheetlike) endothelial cells "glued" together by a polysaccharide (sugar) intercellular matrix, surrounded by a thin layer of subendothelial connective tissue interlaced with a number of circularly arranged elastic fibers to form the subendothelium, and separated from the next adjacent wall layer by a thick elastic band called the *internal elastic lamina,* (2) the thickest *tunica media,* composed of numerous circularly arranged elastic fibers, especially prevalent in the largest blood vessels on the arterial side (allowing them to expand during systole and to recoil passively during diastole), a significant amount of smooth muscle cells arranged in spiraling layers around the vessel wall, especially prevalent in medium-sized arteries and arterioles (allowing them to function as control points for blood distribution), and some interlacing collagenous connective tissue, elastic fibers, and intercellular muco-polysaccharide substance (extractives), all separated from the next adjacent wall layer by another thick elastic band called the *external elastic lamina,* and (3) the medium-sized *tunica adventitia,* an outer vascular sheath consisting entirely of connective tissue.

The largest blood vessels, such as the aorta, the pulmonary artery, the pulmonary veins, and others, have such thick walls that they require a separate network of tiny blood vessels — the vasa vasorum — just to service the vascular tissue itself. As one moves toward the capillaries from the arterial side (see Table 1.3), the vascular wall keeps thinning, as if it were shedding 15-μm-thick, onion-peel-like concentric layers, and while the percentage of water in the vessel wall stays relatively constant at 70% (by weight), the ratio of elastin to collagen decreases (actually reverses) — from 3:2 in large arteries (9% elastin, 6%

TABLE 1.3 Arterial System*

Blood Vessel Type	(Systemic) Typical Number	Internal Diameter Range	Length Range†	Wall Thickness	Systemic Volume	(Pulmonary) Typical Number	Pulmonary Volume
Aorta	1	1.0–3.0 cm	30–65 cm	2–3 mm	156 ml	—	—
Pulmonary artery	—	2.5–3.1 cm	6–9 cm	2–3 cm	—	1	52 ml

Wall morphology: Complete tunica adventitia, external elastic lamina, tunica media, internal elastic lamina, tunica intima, subendothelium, endothelium, and vasa vasorum vascular supply

Blood Vessel Type	(Systemic) Typical Number	Internal Diameter Range	Length Range†	Wall Thickness	Systemic Volume	(Pulmonary) Typical Number	Pulmonary Volume
Main branches	32	5 mm–2.25 cm	3.3–6 cm	≈2 mm	83.2 ml	6	41.6 ml

(Along with the aorta and pulmonary artery, the largest, most well-developed of all blood vessels)

Blood Vessel Type	(Systemic) Typical Number	Internal Diameter Range	Length Range†	Wall Thickness	Systemic Volume	(Pulmonary) Typical Number	Pulmonary Volume
Large arteries	288	4.0–5.0 mm	1.4–2.8 cm	≈1 mm	104 ml	64	23.5 ml

(A well-developed tunica adventitia and vasa vasorum, although wall layers are gradually thinning)

Blood Vessel Type	(Systemic) Typical Number	Internal Diameter Range	Length Range†	Wall Thickness	Systemic Volume	(Pulmonary) Typical Number	Pulmonary Volume
Medium arteries	1152	2.5–4.0 mm	1.0–2.2 cm	≈0.75 mm	117 ml	144	7.3 ml
Small arteries	3456	1.0–2.5 mm	0.6–1.7 cm	≈0.50 mm	104 ml	432	5.7 ml
Tributaries	20,736	0.5–1.0 mm	0.3–1.3 cm	≈0.25 mm	91 ml	5184	7.3 ml

(Well-developed tunica media and external elastic lamina, but tunica adventitia virtually nonexistent)

Blood Vessel Type	(Systemic) Typical Number	Internal Diameter Range	Length Range†	Wall Thickness	Systemic Volume	(Pulmonary) Typical Number	Pulmonary Volume
Small rami	82,944	250–500 μm	0.2–0.8 cm	≈125 μm	57.2 ml	11,664	2.3 ml
Terminal branches	497,664	100–250 μm	1.0–6.0 mm	≈60 μm	52 ml	139,968	3.0 ml

(A well-developed endothelium, subendothelium, and internal elastic lamina, plus about two to three 15-μm-thick concentric layers forming just a very thin tunica media; no external elastic lamina)

Blood Vessel Type	(Systemic) Typical Number	Internal Diameter Range	Length Range†	Wall Thickness	Systemic Volume	(Pulmonary) Typical Number	Pulmonary Volume
Arterioles	18,579,456	25–100 μm	0.2–3.8 mm	≈20–30 μm	52 ml	4,094,064	2.3 ml

Wall morphology: More than one smooth muscle layer (with nerve association in the outermost muscle layer), a well-developed internal elastic lamina; gradually thinning in 25- to 50-μm vessels to a single layer of smooth muscle tissue, connective tissue, and scant supporting tissue.

Blood Vessel Type	(Systemic) Typical Number	Internal Diameter Range	Length Range†	Wall Thickness	Systemic Volume	(Pulmonary) Typical Number	Pulmonary Volume
Metarterioles	238,878,720	10–25 μm	0.1–1.8 mm	≈5–15 μm	41.6 ml	157,306,536	4.0 ml

(Well-developed subendothelium; discontinuous contractile muscle elements; one layer of connective tissue)

Blood Vessel Type	(Systemic) Typical Number	Internal Diameter Range	Length Range†	Wall Thickness	Systemic Volume	(Pulmonary) Typical Number	Pulmonary Volume
Capillaries	16,124,431,360	3.5–10 μm	0.5–1.1 mm	≈0.5–1 μm	260 ml	3,218,406,696	104 ml

(Simple endothelial tubes devoid of smooth muscle tissue; one-cell-layer-thick walls)

*Values are approximate for a 68.7-kg individual having a total blood volume of 5200 ml.
†Average uninterrupted distance between branch origins (except aorta and pulmonary artery, which are total length).

TABLE 1.4 Venous System

Blood Vessel Type	(Systemic) Typical Number	Internal Diameter Range	Length Range	Wall Thickness	Systemic Volume	(Pulmonary) Typical Number	Pulmonary Volume
Postcapillary venules	4,408,161,734	8–30 μm	0.1–0.6 mm	1.0–5.0 μm	166.7 ml	306,110,016	10.4 ml
(Wall consists of thin endothelium exhibiting occasional pericytes (pericapillary connective tissue cells) that increase in number as the vessel lumen gradually increases)							
Collecting venules	160,444,500	30–50 μm	0.1–0.8 mm	5.0–10 μm	161.3 ml	8,503,056	1.2 ml
(One complete layer of pericytes, one complete layer of veil cells (veil-like cells forming a thin membrane), occasional primitive smooth muscle tissue fibers that increase in number with vessel size)							
Muscular venules	32,088,900	50–100 μm	0.2–1.0 mm	10–25 μm	141.8 ml	3,779,136	3.7 ml
(Relatively thick wall of smooth muscle tissue)							
Small collecting veins	10,241,508	100–200 μm	0.5–3.2 mm	≈30 μm	329.6 ml	419,904	6.7 ml
(Prominent tunica media of continuous layers of smooth muscle cells)							
Terminal branches	496,900	200–600 μm	1.0–6.0 mm	30–150 μm	206.6 ml	34,992	5.2 ml
(A well-developed endothelium, subendothelium, and internal elastic lamina; well-developed tunica media but fewer elastic fibers than corresponding arteries and much thinner walls)							
Small veins	19,968	600 μm–1.1 mm	2.0–9.0 mm	≈0.25 mm	63.5 ml	17,280	44.9 ml
Medium veins	512	1–5 mm	1–2 cm	≈0.50 mm	67.0 ml	144	22.0 ml
Large veins	256	5–9 mm	1.4–3.7 cm	≈0.75 mm	476.1 ml	48	29.5 ml
(Well-developed wall layers comparable to large arteries but about 25% thinner)							
Main branches	224	9.0 mm–2.0 cm	2.0–10 cm	≈1.00 mm	1538.1 ml	16	39.4 ml
(Along with the vena cava and pulmonary veins, the largest, most well-developed of all blood vessels)							
Vena cava	1	2.0–3.5 cm	20–50 cm	≈1.50 mm	125.3 ml	—	—
Pulmonary veins	—	1.7–2.5 cm	5–8 cm	≈1.50 mm	—	4	52 ml

Wall morphology: Essentially the same as comparable major arteries but a much thinner tunica intima, a much thinner tunica media, and a somewhat thicker tunica adventitia; contains a vasa vasorum

Total systemic blood volume: 4394 ml — 84.5% of total blood volume; 19.5% in arteries (~3:2 large:small), 5.9% in capillaries, 74.6% in veins (~3:1 large:small); 63% of volume is in vessels greater than 1 mm internal diameter

Total pulmonary blood volume: 468 ml — 9.0% of total blood volume; 31.8% in arteries, 22.2% in capillaries, 46% in veins; 58.3% of volume is in vessels greater than 1 mm internal diameter; remainder of blood in heart, about 338 ml (6.5% of total blood volume)

collagen, by weight) to 1:2 in small tributaries (5% elastin, 10% collagen) — and the amount of smooth muscle tissue increases from 7.5% by weight of large arteries (the remaining 7.5% consisting of various extractives) to 15% in small tributaries. By the time one reaches the capillaries, one encounters single-cell-thick endothelial tubes — devoid of any smooth muscle tissue, elastin, or collagen — downstream of which the vascular wall gradually "reassembles itself," layer by layer, as it directs blood back to the heart through the venous system (Table 1.4).

Blood vessel structure is directly related to function. The thick-walled large arteries and main *distributing branches* are designed to withstand the pulsating 80-to-130-mmHg blood pressures that they must endure. The smaller elastic *conducting vessels* need only operate under steadier blood pressures in the range 70 to 90 mmHg, but they must be thin enough to penetrate and course through organs without unduly disturbing the anatomic integrity of the mass involved. Controlling arterioles operate at blood pressures between 45 and 70 mmHg but are heavily endowed with smooth muscle tissue (hence their referred to as *muscular vessels*) so that they may be actively shut down when flow to the capillary bed they service is to be restricted (for whatever reason), and the smallest capillary *resistance vessels* (which operate at blood pressures on the order of 10 to 45 mmHg) are designed to optimize conditions for transport to occur between blood and the surrounding interstitial fluid. Traveling back up the venous side, one encounters relatively steady blood pressures continuously decreasing from around 30 mmHg all the way down to near zero, so these vessels can be thin-walled without disease consequence. However, the low blood pressure, slower, steady (time-dependent) flow, thin walls, and larger caliber that characterize the venous system cause blood to tend to "pool" in veins, allowing them to act somewhat like reservoirs. It is not surprising, then, that at any given instant, one normally finds about two-thirds of the total human blood volume residing in the venous system, the remaining one-third being divided among the heart (6.5%), the microcirculation (7% in systemic and pulmonary capillaries), and the arterial system (19.5 to 20%).

In a global sense, then, one can think of the human cardiovascular system — using an electrical analogy — as a voltage source (the heart), two capacitors (a large venous system and a smaller arterial system), and a resistor (the microcirculation taken as a whole). Blood flow and the dynamics of the system represent electrical inductance (inertia), and useful engineering approximations can be derived from such a simple model. The cardiovascular system is designed to bring blood to within a capillary size of each and every one of the more than 10^{14} cells of the body — but *which* cells receive blood at any given time, *how much* blood they get, the *composition* of the fluid coursing by them, and related physiologic considerations are all matters that are not left up to chance.

1.4 Cardiovascular Control

Blood flows through organs and tissues either to nourish and sanitize them or to be itself processed in some sense — e.g., to be oxygenated (pulmonary circulation), stocked with nutrients (splanchnic circulation), dialyzed (renal circulation), cooled (cutaneous circulation), filtered of dilapidated red blood cells (splenic circulation), and so on. Thus any given vascular network normally receives blood according to the metabolic needs of the region it perfuses and/or the function of that region as a blood treatment plant and/or thermoregulatory pathway. However, it is not feasible to expect that our physiologic transport system can be "all things to all cells all of the time" — especially when resources are scarce and/or time is a factor. Thus the distribution of blood is further prioritized according to three basic criteria: (1) how essential the perfused region is to the maintenance of life itself (e.g., we can survive without an arm, a leg, a stomach, or even a large portion of our small intestine but not without a brain, a heart, and at least one functioning kidney and lung, (2) how essential the perfused region is in allowing the organism to respond to a life-threatening situation (e.g., digesting a meal is among the least of the body's concerns in a "fight-or-flight" circumstance), and (3) how well the perfused region can function and survive on a decreased supply of blood (e.g., some tissues — like striated skeletal and smooth muscle — have significant anaerobic capability; others — like several forms of connective tissue — can function quite effectively at a significantly decreased metabolic rate when necessary; some organs — like the liver — are larger than they really need to be; and some anatomic structures — like the eyes, ears, and limbs — have duplicates, giving them a built-in redundancy).

Within this generalized prioritization scheme, control of cardiovascular function is accomplished by mechanisms that are based either on the inherent physicochemical attributes of the tissues and organs themselves — so-called intrinsic control — or on responses that can be attributed to the effects on cardiovascular tissues of other organ systems in the body (most notably the autonomic nervous system and the endocrine system) — so-called extrinsic control. For example, the accumulation of wastes and depletion of oxygen and nutrients that accompany the increased rate of metabolism in an active tissue both lead to an *intrinsic* relaxation of local precapillary sphincters (rings of muscle) — with a consequent widening of corresponding capillary entrances — which reduces the local resistance to flow and thereby allows more blood to perfuse the active region. On the other hand, the *extrinsic* innervation by the autonomic nervous system of smooth muscle tissues in the walls of arterioles allows the central nervous system to completely shut down the flow to entire vascular beds (such as the cutaneous circulation) when this becomes necessary (such as during exposure to extremely cold environments).

In addition to prioritizing and controlling the *distribution* of blood, physiologic regulation of cardiovascular function is directed mainly at four other variables: cardiac output, blood pressure, blood volume, and blood composition. From Equation (1.1) we see that cardiac output can be increased by increasing the heart rate (a chronotropic effect), increasing the end-diastolic volume (allowing the heart to fill longer by delaying the onset of systole), decreasing the end-systolic volume (an inotropic effect), or doing all three things at once. Indeed, under the extrinsic influence of the sympathetic nervous system and the adrenal glands, HR can triple — to some 240 beats/min if necessary — EDV can increase by as much as 50% — to around 200 ml or more of blood — and ESV and decrease a comparable amount (the cardiac reserve) — to about 30 to 35 ml or less. The combined result of all three effects can lead to over a sevenfold increase in cardiac output — from the normal 5 to 5.5 liters/min to as much as 40 to 41 liters/min or more for very brief periods of strenuous exertion.

The control of blood pressure is accomplished mainly by adjusting at the arteriolar level the downstream resistance to flow — an increased resistance leading to a rise in arterial backpressure, and vice versa. This effect is conveniently quantified by a fluid-dynamic analogue to Ohm's famous $E = IR$ law in electromagnetic theory, voltage drop E being equated to fluid pressure drop ΔP, electric current I corresponding to flow — cardiac output (CO) — and electric resistance R being associated with an analogous vascular "peripheral resistance" (PR). Thus one may write

$$\Delta P = \left(CO\right)\left(PR\right). \tag{1.4}$$

Normally, the total systemic peripheral resistance is 15 to 20 mmHg/liter/min of flow but can increase significantly under the influence of the vasomotor center located in the medulla of the brain, which controls arteriolar muscle tone.

The control of blood volume is accomplished mainly through the excretory function of the kidney. For example, antidiuretic hormone (ADH) secreted by the pituitary gland acts to prevent renal fluid loss (excretion via urination) and thus increases plasma volume, whereas perceived extracellular fluid overloads such as those which result from the peripheral vasoconstriction response to cold stress lead to a sympathetic/adrenergic receptor-induced renal diuresis (urination) that tends to decrease plasma volume — if not checked, to sometimes dangerously low dehydration levels. Blood composition, too, is maintained primarily through the activity of endocrine hormones and enzymes that enhance or repress specific biochemical pathways. Since these pathways are too numerous to itemize here, suffice it to say that in the body's quest for homeostasis and stability, virtually nothing is left to chance, and every biochemical end can be arrived at through a number of alternative means. In a broader sense, as the organism strives to maintain life, it coordinates a wide variety of different functions, and central to its ability to do just that is the role played by the cardiovascular system in transporting mass, energy, and momentum.

Defining Terms

Atrioventricular (AV) node: A highly specialized cluster of neuromuscular cells at the lower portion of the right atrium leading to the interventricular septum; the AV node delays sinoatrial, (SA) node–generated electrical impulses momentarily (allowing the atria to contract first) and then conducts the depolarization wave to the bundle of His and its bundle branches.

Autonomic nervous system: The functional division of the nervous system that innervates most glands, the heart, and smooth muscle tissue in order to maintain the internal environment of the body.

Cardiac muscle: Involuntary muscle possessing much of the anatomic attributes of skeletal voluntary muscle and some of the physiologic attributes of involuntary smooth muscle tissue; SA node–induced contraction of its interconnected network of fibers allows the heart to expel blood during systole.

Chronotropic: Affecting the periodicity of a recurring action, such as the slowing (bradycardia) or speeding up (tachycardia) of the heartbeat that results from extrinsic control of the SA node.

Endocrine system: The system of ductless glands and organs secreting substances directly into the blood to produce a specific response from another "target" organ or body part.

Endothelium: Flat cells that line the innermost surfaces of blood and lymphatic vessels and the heart.

Homeostasis: A tendency to uniformity or stability in an organism by maintaining within narrow limits certain variables that are critical to life.

Inotropic: Affecting the contractility of muscular tissue, such as the increase in cardiac *power* that results from extrinsic control of the myocardial musculature.

Precapillary sphincters: Rings of smooth muscle surrounding the entrance to capillaries where they branch off from upstream metarterioles. Contraction and relaxation of these sphincters close and open the access to downstream blood vessels, thus controlling the irrigation of different capillary networks.

Sinoatrial (SA) node: Neuromuscular tissue in the right atrium near where the superior vena cava joins the posterior right atrium (the sinus venarum); the SA node generates electrical impulses that initiate the heartbeat, hence its nickname the cardiac "pacemaker."

Stem cells: A generalized parent cell spawning descendants that become individually specialized.

Acknowledgments

The author gratefully acknowledges the assistance of Professor Robert Hochmuth in the preparation of Table 1.1 and the Radford Community Hospital for their support of the Biomedical Engineering Program at Virginia Tech.

References

Bhagavan NV. 1992. *Medical Biochemistry.* Boston, Jones and Bartlett.

Beall HPT, Needham D, Hochmuth RM. 1993. Volume and osmotic properties of human neutrophils. *Blood* 81(10):2774–2780.

Caro CG, Pedley TJ, Schroter RC, Seed WA. 1978. *The Mechanics of the Circulation.* New York, Oxford University Press.

Chandran KB. 1992. *Cardiovascular Biomechanics.* New York, New York University Press.

Frausto da Silva JJR, Williams RJP. 1993. *The Biological Chemistry of the Elements.* New York, Oxford University Press/Clarendon.

Dawson TH. 1991. *Engineering Design of the Cardiovascular System of Mammals.* Englewood Cliffs, NJ, Prentice-Hall.

Duck FA. 1990. *Physical Properties of Tissue.* San Diego, Academic Press.

Kaley G, Altura BM (Eds). *Microcirculation, Vol I (1977), Vol II (1978), Vol III (1980).* Baltimore, University Park Press.

Kessel RG, Kardon RH. 1979. *Tissue and Organs — A Text-Atlas of Scanning Electron Microscopy.* San Francisco, WH Freeman.

Lentner C (Ed). *Geigy Scientific Tables, Vol 3: Physical Chemistry, Composition of Blood, Hematology and Somatometric Data*, 8th ed. 1984. New Jersey, Ciba-Geigy.

— — — *Vol 5: Heart and Circulation*, 8th ed. 1990. New Jersey, Ciba-Geigy.

Schneck DJ. 1990. *Engineering Principles of Physiologic Function.* New York, New York University Press.

Tortora GJ, Grabowski SR. 1993. *Principles of Anatomy and Physiology*, 7th ed. New York, HarperCollins.

2
Endocrine System

Derek G. Cramp
City University, London

Ewart R. Carson
City University, London

The body, if it is to achieve optimal performance, must possess mechanisms for sensing and responding appropriately to numerous biologic cues and signals in order to control and maintain its internal environment. This complex role is effected by the integrative action of the endocrine and neural systems. The endocrine contribution is achieved through a highly sophisticated set of communication and control systems involving signal generation, propagation, recognition, transduction, and response. The signal entities are chemical messengers or hormones that are distributed through the body by the blood circulatory system to their respective target organs to modify their activity in some fashion.

Endocrinology has a comparatively long history, but real advances in the understanding of endocrine physiology and mechanisms of regulation and control only began in the late 1960s with the introduction of sensitive and relatively specific analytical methods; these enabled low concentrations of circulating hormones to be measured reliably, simply, and at relatively low cost. The breakthrough came with the development and widespread adoption of competitive protein binding and radioimmunoassays that superseded existing cumbersome bioassay methods. Since then, knowledge of the physiology of individual endocrine glands and of the neural control of the pituitary gland and the overall feedback control of the endocrine system has progressed and is growing rapidly. Much of this has been accomplished by applying to endocrinological research the methods developed in cellular and molecular biology and recombinant DNA technology. At the same time, theoretical and quantitative approaches using mathematical modeling complemented experimental studies have been of value in gaining a greater understanding of endocrine dynamics.

2.1   Endocrine System: Hormones, Signals, and Communication Between Cells and Tissues

Hormones are synthesized and secreted by specialized endocrine glands to act locally or at a distance, having been carried in the bloodstream (classic endocrine activity) or secreted into the gut lumen (lumocrine activity) to act on target cells that are distributed elsewhere in the body. Hormones are chemically diverse, physiologically potent molecules that are the primary vehicle for intercellular communication with the

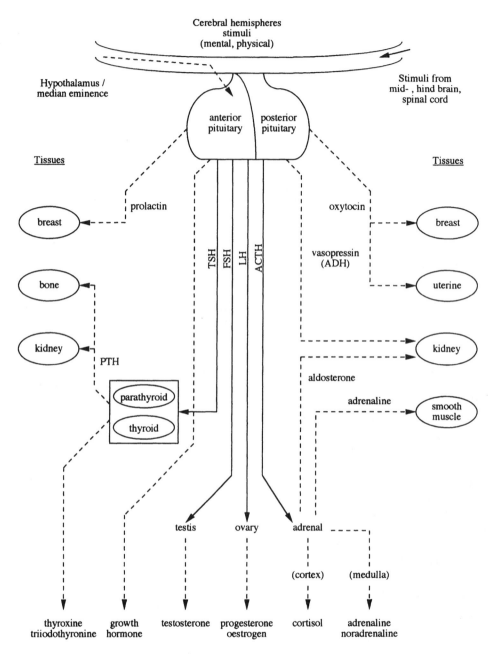

**FIGURE 2.1   Representation of the forward pathways of pituitary and target gland hormone release and action:
— — tropic hormones; – – – tissue-affecting hormones.**

capacity to override the intrinsic mechanisms of normal cellular control. They can be classified broadly
into three groups according to their physicochemical characteristics: (1) steroid hormones produced by
chemical modification of cholesterol, (2) peptide and protein hormones, and (3) those derived from the
aromatic amino acid tyrosine. The peptide and protein hormones are essentially hydrophilic and are
therefore able to circulate in the blood in the free state; however, the more hydrophobic lipid-derived
molecules have to be carried in the circulation bound to specific transport proteins. Figure 2.1 and Table
2.1 show, in schematic and descriptive form respectively, details of the major endocrine glands of the
body and the endocrine pathways.

TABLE 2.1 Main Endocrine Glands and the Hormones They Produce and Release

Gland	Hormone	Chemical Characteristics
Hypothalamus/median eminence	Thyrotropin-releasing hormone (TRH)	Peptides
	Somatostatin	
	Gonadotropin-releasing hormone	Amine
	Growth hormone-releasing hormone	
	Corticotropin-releasing hormone	
	Prolactin inhibitor factor	
Anterior pituitary	Thyrotropin (TSH)	Glycoproteins
	Luteinizing hormone	
	Follicle-stimulating hormone (FSH)	Proteins
	Growth hormone	
	Prolactin	
	Adrenocorticotropin (ACTH)	
Posterior pituitary	Vasopressin (antidiuretic hormone, ADH)	
	Oxytocin	Peptides
Thyroid	Triidothyronine (T3)	Tyrosine derivatives
	Thyroxine (T4)	
Parathyroid	Parathyroid hormone (PTH)	Peptide
Adrenal cortex	Cortisol	Steroids
	Aldosterone	
Adrenal medulla	Epinephrine	Catecolamines
	Norepinephrine	
Pancreas	Insulin	Proteins
	Glucagon	
	Somatostatin	
Gonads:Testes	Testosterone	Steroids
Ovaries	Estrogen	
	Progesterone	

The endocrine and nervous system are physically and functionally linked by a specific region of the brain called the *hypothalamus*, which lies immediately above the pituitary gland, to which it is connected by an extension called the *pituitary stalk*. The integrating function of the hypothalamus is mediated by cells that possess the properties of both nerve and processes that carry electrical impulses and on stimulation can release their signal molecules into the blood. Each of the hypothalamic neurosecretory cells can be stimulated by other nerve cells in higher regions of the brain to secrete specific peptide hormones or release factors into the adenohypophyseal portal vasculature. These hormones can then specifically stimulate or suppress the secretion of a second hormone from the anterior pituitary.

The pituitary hormones in the circulation interact with their target tissues, which, if endocrine glands, are stimulated to secrete further (third) hormones that feed back to inhibit the release of the pituitary hormones. It will be seen from Fig. 2.1 and Table 2.1 that the main targets of the pituitary are the adrenal cortex, the thyroid, and the gonads. These axes provide good examples of the control of pituitary hormone release by negative-feedback inhibition; e.g., adrenocorticotropin (ACTH), luteinizing hormone (LH), and follicle-stimulating hormone (FSH) are selectively inhibited by different steroid hormones, as is thyrotropin (TSH) release by the thyroid hormones.

In the case of growth hormone (GH) and prolactin, the target tissue is not an endocrine gland and thus does not produce a hormone; then the feedback control is mediated by inhibitors. Prolactin is under dopamine inhibitory control, whereas hypothalamic releasing and inhibitory factors control GH release. The two posterior pituitary (neurohypophyseal) hormones, oxytocin and vasopressin, are synthesized in the supraoptic and paraventricular nuclei and are stored in granules at the end of the nerve fibers in the posterior pituitary. Oxytocin is subsequently secreted in response to peripheral stimuli from the cervical stretch receptors or the suckling receptors of the breast. In a like manner, antidiuretic hormone (ADH, vasopressin) release is stimulated by the altered activity of hypothalamic osmoreceptors responding to changes in plasma solute concentrations.

It will be noted that the whole system is composed of several endocrine axes with the hypothalamus, pituitary, and other endocrine glands together forming a complex hierarchical regulatory system. There is no doubt that the anterior pituitary occupies a central position in the control of hormone secretion and, because of its important role, was often called the "conductor of the endocrine orchestra." However, the release of pituitary hormones is mediated by complex feedback control, so the pituitary should be regarded as having a permissive role rather than having the overall control of the endocrine system.

2.2 Hormone Action at the Cell Level: Signal Recognition, Signal Transduction, and Effecting a Physiological Response

The ability of target glands or tissues to respond to hormonal signals depends on the ability of the cells to recognize the signal. This function is mediated by specialized proteins or glycoproteins in or on the cell plasma membrane that are specific for a particular hormone, able to recognize it, bind it with high affinity, and react when very low concentrations are present. Recognition of the hormonal signal and activation of the cell surface receptors initiates a flow of information to the cell interior that triggers a chain of intracellular events in a preprogrammed fashion that produces a characteristic response. It is useful to classify the site of such action of hormones into two groups: those that act at the cell surface without, generally, traversing the cell membrane and those that actually enter the cell before effecting a response. In the study of this multistep sequence, two important events can be readily studied, namely, the binding of the hormone to its receptor and activation of cytoplasmic effects. However, it is some of the steps between these events, such as receptor activation and signal generation, that are still relatively poorly defined. One method employed in an attempt to elucidate the intermediate steps has been to use ineffective mutant receptors, which when assayed are either defective in their hormone-binding capabilities or in effector-activation and thus unable to transduce a meaningful signal to the cell. But, the difficulty with these studies has been to distinguish receptor-activation and signal-generation defects from hormone-binding and effector-activation defects.

Hormones Acting at the Cell Surface

Most peptide and protein hormones are hydrophilic and thus unable to traverse the lipid-containing cell membrane and must therefore act through activation of receptor proteins on the cell surface. When these receptors are activated by the binding of an extracellular signal ligand, the ligand-receptor complex initiates a series of protein interactions within or adjacent to the inner surface of the plasma membrane, which in turn brings about changes in intracellular activity. This can happen in one of two ways. The first involves the so-called second messenger, by altering the activity of a plasma membrane-bound enzyme, which in turn increases (or sometimes decreases) the concentration of an intracellular mediator. The second involves activation of other types of cell surface receptors, which leads to changes in the plasma membrane electrical potential and the membrane permeability, resulting in altered transmembrane transport of ions or metabolites. If the hormone is thought of as the "first messenger," cyclic adenosine monophosphate (cAMP) can be regarded as the "second messenger"; capable of triggering a cascade of intracellular biochemical events that can lead either to a rapid secondary response, such as altered ion transport, enhanced metabolic pathway flux, steroidogenesis or to a slower response, such as DNA, RNA, and protein synthesis resulting in cell growth or cell division.

The peptide and protein hormones circulate at very low concentrations relative to other proteins in the blood plasma. These low concentrations are reflected in the very high affinity and specificity of the receptor sites, which permits recognition of the relevant hormones amid the profusion of protein molecules in the circulation. Adaptation to a high concentration of a signal ligand in a time-dependent reversible manner enables cells to respond to changes in the concentration of a ligand instead of to its absolute concentration. The number of receptors in a cell is not constant; synthesis of receptors may be induced or repressed by other hormones or even by their own hormones. Adaptation can occur in several ways. Ligand binding can inactivate a cell surface receptor either by inducing its internalization and degradation or by causing the receptor to adopt an inactive conformation. Alternatively, it may result from the changes in one of the nonreceptor proteins involved in signal transduction following receptor activation. Downregulation is the name given to the process whereby a cell decreases the number of receptors in response to intense or frequent stimulation and can occur by degradation or more temporarily by phosphorylation and sequestration. Upregulation is the process of increasing receptor expression either by other hormones or in response to altered stimulation.

The cell surface receptors for peptide hormones are linked functionally to a cell membrane-bound enzyme that acts as the catalytic unit. This receptor complex consists of three components: (1) the receptor itself that recognizes the hormone, (2) a regulatory protein called a G-protein that binds guanine nucleotides and is located on the cytosolic face of the membrane, and (3) adenylate cyclase, which catalyzes the conversion of ATP to cyclic AMP. As the hormone binds at the receptor site, it is coupled through a regulatory protein, which acts as a transducer, to the enzyme adenyl cyclase, which catalyzes the formation of cAMP from adenosine triphosphate (ATP). The G-protein consists of 3 subunits, which in the unstimulated state form a heterotrimer to which a molecule of GDP is bound. Binding of the hormone to the receptor causes the subunit to exchange its GDP for a molecule of GTP (guanine triphosphate), which then dissociates from the subunits. This in turn decreases the affinity of the receptor for the hormone and leads to its dissociation. The GTP subunit not only activates adenylate cyclase, but also has intrinsic GTPase activity and slowly converts GTP back to GDP, thus allowing the subunits to reassociate and so regain their initial resting state. There are hormones, such as somatostatin, that possess the ability to inhibit AMP formation but still have similarly structured receptor complexes. The G-protein of inhibitory complexes consists of an inhibitory subunit complexed with a subunit thought to be identical to the subunits of the stimulatory G-protein. But, it appears that a single adenylate cyclase molecule can be simultaneously regulated by more than one G-protein enabling the system to integrate opposing inputs.

The adenylate cyclase reaction is rapid, and the increased concentration of intracellular cAMP is short-lived, since it is rapidly hydrolyzed and destroyed by the enzyme cAMP phosphodiesterase that terminates the hormonal response. The continual and rapid removal of cAMP and free calcium ions from the cytosol makes for both the rapid increase and decrease of these intracellular mediators when the cells respond to signals. Rising cAMP concentrations affect cells by stimulating cAMP-dependent protein kinases to phosphorylate specific target proteins. Phosphorylation of proteins leads to conformational changes that enhance their catalytic activity, thus providing a signal amplification pathway from hormone to effector. These effects are reversible because phosphorylated proteins are rapidly dephosphorylated by protein phosphatases when the concentration of cAMP falls. A similar system involving cyclic GMP, although less common and less well studied, plays an analogous role to that of cAMP. The action of thyrotropin-releasing hormone (TRH), parathyroid hormone (PTH), and epinephrine is catalyzed by adenyl cyclase, and this can be regarded as the classic reaction.

However, there are variant mechanisms. In the phosphatidylinositol-diacylglycerol (DAG)/inositol triphosphate (IP3) system, some surface receptors are coupled through another G-protein to the enzyme phospholipase C, which cleaves the membrane phospholipid to form DAG and IP3 or phospholipase D, which cleaves phosphatidyl choline to DAG via phosphatidic acid. DAG causes the calcium, phospholid-dependent protein kinase C to translocate to the the cell membrane from the cytosolic cell compartment, becoming 20 times more active in the process. IP3 mobilizes calcium from storage sites associated with the plasma and intracellular membranes thereby contributing to the activation of protein kinase C as well as other calcium dependent processes. DAG is cleared from the cell either by conversion to phosphatidic acid which may be recycled to phospholipid or it may be broken down to fatty acids and glycerol. The DAG derived from phosphatidylinositol usually contains arachidonic acid esterified to the middle carbon of glycerol. Arachidonic acid is the precursor of the prostaglandins and leukotrienes that are biologically active eicosanoids.

Thyrotropin and vasopressin modulate an activity of phospholipase C that catalyzes the conversion of phosphatidylinositol to diacylglycerol and inositol, l, 4, 5-triphosphate, which act as the second messengers. They mobilize bound intracellular calcium and activate a protein kinase, which in turn alters the activity of other calcium-dependent enzymes within the cell.

Increased concentrations of free calcium ions affect cellular events by binding to and altering the molecular conformation of calmodulin; the resulting calcium ion-calmodulin complex can activate many different target proteins, including calcium ion-dependent protein kinases. Each cell type has a characteristic set of target proteins that are so regulated by cAMP-dependent kinases and/or calmodulin that it will respond in a specific way to an alteration in cAMP or calcium ion concentrations. In this way, cAMP or calcium ions act as second messengers in such a way as to allow the extracellular signal not only to be greatly amplified but, just as importantly, also to be made specific for each cell type.

The action of the important hormone insulin, which regulates glucose metabolism, depends on the activation of the enzyme tyrosine kinase catalyzing the phosphorylation of tyrosyl residues of proteins. This effects changes in the activity of calcium-sensitive enzymes, leading to enhanced movement of glucose and fatty acids across the cell membrane and modulating their intracellular metabolism. The binding of insulin to its receptor site has been studied extensively; the receptor complex has been isolated and characterized. It was such work that highlighted the interesting aspect of feedback control at the cell level, downregulation: the ability of peptide hormones to regulate the concentration of cell surface receptors. After activation, the receptor population becomes desensitized, or "downregulated"; leading to a decreased availability of receptors and thus a modulation of transmembrane events.

Hormones Acting Within the Cell

Steroid hormones are small hydrophobic molecules derived from cholesterol that are solubilized by binding reversibly to specify carrier proteins in the blood plasma. Once released from their carrier proteins, they readily pass through the plasma membrane of the target cell and bind, again reversibly, to steroid hormone receptor proteins in the cytosol. This is a relatively slow process when compared to protein hormones. The latter second messenger-mediated phosphorylation-dephosphorylation reactions modify enzymatic processes rapidly with the physiological consequences becoming apparent in seconds or minutes and are as rapidly reversed. Nuclear-mediated responses, on the other hand, lead to transcription/translation dependent changes that are slow in onset and tend to persist since reversal is dependent on degradation of the induced proteins. The protein component of the steroid hormone-receptor complex has an affinity for DNA in the cell nucleus, where it binds to nuclear chromatin and initiates the transcription of a small number of genes. These gene products may, in turn, activate other genes and produce a secondary response, thereby amplifying the initial effect of the hormone. Each steroid hormone is recognized by a separate receptor protein, but this same receptor protein has the capacity to regulate several different genes in different target cells. This, of course, suggests that the nuclear chromatin of each cell type is organized so that only the appropriate genes are made available for regulation by the hormone-receptor complex. The thyroid hormone triiodothyronine (T3) also acts, though by a different mechanism than the steroids, at the cell nucleus level to initiate genomic transcription. The hormonal activities of GH and prolactin influence cellular gene transcription and translation of messenger RNA by complex mechanisms.

2.3 Endocrine System: Some Other Aspects of Regulation and Control

From the foregoing sections it is clear that the endocrine system exhibits complex molecular and metabolic dynamics that involve many levels of control and regulation. Hormones are chemical signals released from a hierarchy of endocrine glands and propagated through the circulation to a hierarchy of cell types. The integration of this system depends on a series of what systems engineers call "feedback loops"; feedback is a reflection of mutual dependence of the system variables: variable

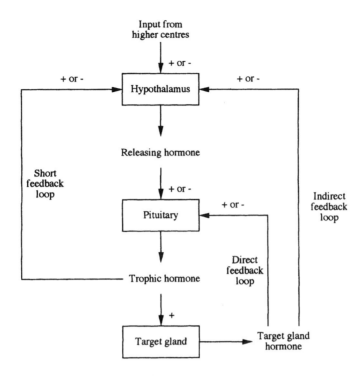

FIGURE 2.2   Illustration of the complexity of hormonal feedback control (+ indicates a positive or augmenting effect; – indicates a negative or inhibiting effect).

x affects variable *y*, and *y* affects *x*. Further, it is essentially a closed-loop system in which the feedback of information from the system output to the input has the capacity to maintain homeostasis. A diagrammatic representation of the ways in which hormone action is controlled is shown in Fig. 2.2. One example of this control structure arises in the context of the thyroid hormones. In this case, TRH, secreted by the hypothalamus, triggers the anterior pituitary into the production of TSH. The target gland is the thyroid, which produces T3 and thyroxine (T4). The complexity of control includes both direct and indirect feedback of T3 and T4, as outlined in Fig. 2.2, together with TSH feedback on to the hypothalamus.

Negative Feedback

If an increase in *y* causes a change in *x*, which in turn tends to decrease *y*, feedback is said to be *negative;* in other words, the signal output induces a response that feeds back to the signal generator to decrease its output. This is the most common form of control in physiologic systems, and examples are many. For instance, as mentioned earlier, the anterior pituitary releases trophic or stimulating hormones that act on peripheral endocrine glands, such as the adrenals or thyroid or to gonads, to produce hormones that act back on the pituitary to decrease the secretion of the trophic hormones. These are examples of what is called *long-loop feedback (see* Fig. 2.2). (**Note:** the adjectives *long* and *short* reflect the spatial distance or proximity of effector and target sites.) The trophic hormones of the pituitary are also regulated by feedback action at the level of their releasing factors. *Ultrashort-loop feedback* is also described. There are numerous examples of *short-loop feedback* as well, the best being the reciprocal relation between insulin and blood glucose concentrations, as depicted in Fig. 2.3. In this case, elevated glucose concentration (and positive rate of change, implying not only proportional but also derivative control) has a positive effect on the pancreas, which secretes insulin in response. This has an inhibiting effect on glucose metabolism, resulting in a reduction of blood glucose toward a normal concentration; in other words, classic negative-feedback control.

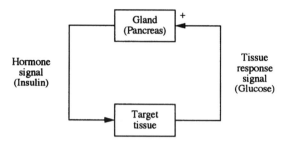

FIGURE 2.3   **The interaction of insulin as an illustration of negative feedback within a hormonal control system.**

Positive Feedback

If increase in *y* causes a change in *x* that tends to increase *y*, feedback is said to be *positive*; in other words, a further signal output is evoked by the response it induces or provokes. This is intrinsically an unstable system, but there are physiologic situations where such control is valuable. In the positive feedback situation, the signal output will continue until no further response is required. Suckling provides an example; stimulation of nipple receptors by the suckling child provokes an increased oxytocin release from the posterior pituitary with a corresponding increase in milk flow. Removal of the stimulus causes cessation of oxytocin release.

Rhythmic Endocrine Control

Many hormone functions exhibit rhythmically in the form of pulsatile release of hormones. The most common is the approximately 24-h cycle (circadian or diurnal rhythm). For instance, blood sampling at frequent intervals has shown that ACTH is secreted episodically, each secretory burst being followed 5 to 10 min later by cortisol secretion. These episodes are most frequent in the early morning, with plasma cortisol concentrations highest around 7 to 8 AM and lowest around midnight. ACTH and cortisol secretion vary inversely, and the parallel circadian rhythm is probably due to a cyclic change in the sensitivity of the hypothalamic feedback center to circulating cortisol. Longer cycles are also known, e.g., the infradian menstrual cycle.

It is clear that such inherent rhythms are important in endocrine communication and control, suggesting that its physiologic organization is based not only on the structural components of the system but also on the dynamics of their interactions. The rhythmic, pulsatile nature of release of many hormones is a means whereby time-varying signals can be encoded, thus allowing large quantities of information to be transmitted and exchanged rapidly in a way that small, continuous changes in threshold levels would not allow.

References

Inevitably, our brief exposition has been able to touch upon an enormous subject only by describing some of the salient features of this fascinating domain, but it is hoped that it may nevertheless stimulate a further interest. However, not surprisingly the endocrinology literature is massive, and it is suggested that anyone wishing to read further go initially to one of the many excellent textbooks and go on from there. Those we have found useful include:

Goodman HM. 1994. *Basic Medical Endocrinology,* 2nd ed., New York, Raven Press.
Greenspan FS, Strewler GJ (Eds.). 1997. *Basic and Clinical Endocrinology,* Appleton and Lange. 5th ed.
O'Malley BW, Birnbaumer L, Hunter T. (Eds.). 1998. *Hormones and Signaling.* (Vol. 1). Academic Press.
Wilson JD, Foster DW, Kronenberg HM (Eds.). 1998. *Williams Textbook of Endocrinology,* 9th ed., WB
    Saunders.

3

Nervous System

Evangelia
Micheli-Tzanakou
Rutgers University

The nervous system, unlike other organ systems, is concerned primarily with signals, information encoding and processing, and control rather than manipulation of energy. It acts like a communication device whose components use substances and energy in processing signals and in reorganizing them, choosing, and commanding, as well as in developing and learning. A central question that is often asked is how nervous systems work and what are the principles of their operation. In an attempt to answer this question, we will, at the same time, ignore other fundamental questions, such as those relating to anatomic or neurochemical and molecular aspects. We will concentrate rather on relations and transactions between neurons and their assemblages in the nervous system. We will deal with neural signals (encoding and decoding), the evaluation and weighting of incoming signals, and the formulation of outputs. A major part of this chapter is devoted to higher aspects of the nervous system, such as memory and learning, rather than individual systems, such as vision and audition, which are treated extensively elsewhere in this handbook.

3.1 Definitions

Nervous systems can be defined as organized assemblies of nerve cells as well as nonnervous cells. Nerve cells, or *neurons,* are specialized in the generation, integration, and conduction of incoming signals from the outside world or from other neurons and deliver them to other excitable cells or to *effectors* such as muscle cells. Nervous systems are easily recognized in higher animals but not in the lower species, since the defining criteria are difficult to apply.

A central nervous system (CNS) can be distinguished easily from a peripheral nervous system (PNS), since it contains most of the motor and nucleated parts of neurons that innervate muscles and other effectors. The PNS contains all the sensory nerve cell bodies, with some exceptions, plus local *plexuses,* local *ganglia,* and peripheral axons that make up the *nerves.* Most sensory axons go all the way into the CNS, while the remaining sensory axons relay in peripheral plexuses. Motor axons originating in the CNS innervate effector cells.

The nervous system has two major roles: (1) to regulate, acting homeostatically in restoring some conditions of the organism after some external stimulus, and (2) to act to alter a preexisting condition by replacing it or modifying it. In both cases — regulation or initiation of a process — learning can be superimposed. In most species, learning is a more or less adaptive mechanism, combining and timing species-characteristic acts with a large degree of evolution toward perfection.

The nervous system is a complex structure for which realistic assumptions have led to irrelevant oversimplifications. The nervous system can be broken down into four components: sensory transducers, neurons, axons, and muscle fibers. Each of these components gathers, processes, and transmits information impinging on it from the outside world, usually in the form of complex stimuli. The processing is carried out by exitable tissues — neurons, axons, sensory receptors, and muscle fibers. Neurons are the basic elements of the nervous system. If put in small assemblies or clusters, they form neuronal assemblies or neuronal networks communicating with each other either chemically via *synaptic junctions* or electrically via *tight* junctions. The main characteristics of a cell are the *cell body,* or *soma,* which contains the *nucleus,* and a number of processes originating from the cell body, called the *dendrites,* which reach out to surroundings to make contacts with other cells. These contacts serve as the incoming information to the cell, while the outgoing information follows a conduction path, the axon. The incoming information is integrated in the cell body and generates its action potential at the *axon hillock.* There are two types of outputs that can be generated and therefore two types of neurons: those that generate *graded* potentials that attenuate with distance and those that generate *action* potentials. The latter travel through the axon, a thin, long process that passively passes the action potential or rather a train of action potentials without any attenuation (*all-or-none effect*). A number of action potentials is often called a *spike train.* A threshold built into the hillock, depending on its level, allows or stops the generation of the spike train. Axons usually terminate on other neurons by means of *synaptic terminals* or *boutons* and have properties similar to those of an electric cable with varying diameters and speeds of signal transmission. Axons can be of two types: *myelinated* or *unmyelinated.* In the former case, the axon is surrounded by a thick fatty material, the myelin sheath, which is interrupted at regular intervals by gaps called the *nodes of Ranvier.* These nodes provide for the *saltatory* conduction of the signal along the axon. The axon makes functional connections with other neurons at synapses on either the cell body, the dendrites, or the axons. There exist two kinds of synapses: *excitatory* and *inhibitory,* and as the names imply, they either increase the *firing* frequency of the postsynaptic neurons or decrease it, respectively.

Sensory receptors are specialized cells that, in response to an incoming stimulus, generate a corresponding electrical signal, a graded receptor potential. Although the mechanisms by which the sensory receptors generate receptor potentials are not known exactly, the most plausible scenario is that an external stimulus alters the membrane permeabilities. The receptor potential, then, is the change in intracellular potential relative to the *resting* potential.

It is important to notice here that the term *receptor* is used in physiology to refer not only to sensory receptors but also, in a different sense, to proteins that bind neurotransmitters, hormones, and other substances with great affinity and specificity as a first step in starting up physiologic responses. This receptor is often associated with nonneural cells that surround it and form a *sense organ.* The forms of energy converted by the receptors include mechanical, thermal, electromagnetic, and chemical energy. The particular form of energy to which a receptor is most sensitive is called its *adequate stimulus.* The problem of how receptors convert energy into action potentials in the sensory nerves has been the subject of intensive study. In the complex sense organs such as those concerned with hearing and vision, there exist separate receptor cells and synaptic junctions between receptors and afferent nerves. In other cases, such as the cutaneous sense organs, the receptors are specialized. Where a stimulus of constant strength is applied to a receptor repeatedly, the frequency of the action potentials in its sensory nerve declines over a period of time. This phenomenon is known as *adaptation*; if the adaptation is very rapid, then the receptors are called *phasic*; otherwise, they are called *tonic.*

Another important issue is the *coding* of sensory information. Action potentials are similar in all nerves, although there are variations in their speed of conduction and other characteristics. However, if the action potentials were the same in most cells, then what makes the visual cells sensitive to light and not to sound and the touch receptors sensitive to touch and not to smell? And how can we tell if these sensations are strong or not? These sensations depend on what is called the *doctrine of specific nerve energies,* which has been questioned over time by several researchers. No matter where a particular sensory pathway is stimulated along its course to the brain, the sensation produced is referred to the location of the receptor. This is the *law of projections*. An example of this law is the "phantom limb," in which an amputee complains about an itching sensation in the amputated limb.

3.2   Functions of the Nervous System

The basic unit of integrated activity is the *reflex arc*. This arc consists of a sense organ, afferent neuron, one or more synapses in a central integrating station (or sympathetic ganglion), an efferent neuron, and an effector. The simplest reflex arc is the *monosynaptic* one, which has only one synapse between the afferent and efferent neurons. With more than one synapse, the reflex arc is called *polysynaptic*. In each of these cases, activity is modified by both spatial and temporal facilitation, occlusion, and other effects.

In mammals, the concentration between afferent and efferent somatic neurons is found either in the brain or in the spinal cord. The Bell-Magendie law dictates that in the spinal cord the dorsal roots are sensory, while the ventral roots are motor. The action potential message that is carried by an axon is eventually fed to a muscle, to a secretory cell, or to the dendrite of another neuron. If an axon is carrying a graded potential, its output is too weak to stimulate a muscle, but it can terminate on a secretory cell or dendrite. The latter can have as many as 10,000 inputs. If the endpoint is a motor neuron, which has been found experimentally in the case of fibers from the primary endings, then there is a lag between the time when the stimulus was applied and when the response is obtained from the muscle. This time interval is called the *reaction time* and in humans is approximately 20 ms for a stretch reflex. The distance from the spinal cord can be measured, and since the conduction velocities of both the efferent and afferent fibers are known, another important quality can be calculated: the *central delay*. This delay is the portion of the reaction time that was spent for conduction to and from the spinal cord. It has been found that muscle spindles also make connections that cause muscle contraction via polysynaptic pathways, while the afferents from secondary endings make connections that excite extensor muscles. When a motor neuron sends a burst of action potentials to its skeletal muscle, the amount of contraction depends largely on the discharge frequency but also on many other factors, such as the history of the load on the muscle and the load itself. The *stretch error* can be calculated from the desired motion minus the actual stretch. If this error is then fed back to the motor neuron, its discharge frequency is modified appropriately. This corresponds to one of the three feedback loops that are available locally. Another loop corrects for overstretching beyond the point that the muscle or tendon may tear. Since a muscle can only contract, it must be paired with another muscle (*antagonist*) in order to effect the return motion. Generally speaking, a flexor muscle is paired with an extensor muscle that cannot be activated simultaneously. This means that the motor neurons that affect each one of these are not activated at the same time. Instead, when one set of motor neurons is activated, the other is inhibited, and vice versa. When movement involves two or more muscles that normally cooperate by contracting simultaneously, the excitation of one causes facilitation of the other *synergistic* members via cross-connections. All these networks form feedback loops. An engineer's interpretation of how these loops work would be to assume dynamic conditions, as is the case in all parts of the nervous system. This has little value in dealing with stationary conditions, but it provides for an ability to adjust to changing conditions.

The nervous system, as mentioned earlier, is a control system of processes that adjust both internal and external operations. As humans, we have experiences that change our perceptions of events in our environment. The same is true for higher animals, which, besides having an internal environment the status of which is of major importance, also share an external environment of utmost richness and variety. Objects and conditions that have direct contact with the surface of an animal directly affect the future of the animal. Information about changes at some point provides a prediction of possible future status. The amount of information required to represent changing conditions increases as the required temporal resolution of detail increases. This creates a vast amount of data to be processed by any finite system. Considering the fact that the information reaching sensory receptors is too extensive and redundant, as well as modified by external interference (noise), the nervous system has a tremendously difficult task to accomplish. Enhanced responsiveness to a particular stimulus can be produced by structures that either increase the energy converging on a receptor or increase the effectiveness of coupling of a specific type of stimulus with its receptor. Different species have sensory systems that respond to stimuli that are important to them for survival. Often one nervous system responds to conditions that are not sensed by another nervous system. The transduction, processing, and transmission of signals in any nervous system produce a survival mechanism for an organism but only after these signals have been further modified by effector organs. Although the nerve impulses that drive a muscle, as explained earlier, are discrete events, a muscle twitch takes much longer to happen, a fact that allows for their responses to overlap and produce a much smoother output. Neural control of motor activity of skeletal muscle is accomplished entirely by the modification of muscle excitation, which involves changes in velocity, length, stiffness, and heat production. The important of accurate timing of inputs and the maintenance of this timing across several synapses is obvious in sensory pathways of the nervous system. Cells are located next to other cells that have overlapping or adjacent receptor or motor fields. The dendrites provide important and complicated sites of interactions as well as channels of variable effectiveness for excitatory inputs, depending on their position relative to the cell body. Among the best examples are the cells of the medial superior olive in the auditory pathway. These cells have two major dendritic trees extending from opposite poles of the cell body. One receives synaptic inhibitory input from the ipsilateral cochlear nucleus, the other from the contralateral nucleus that normally is an excitatory input. These cells deal with the determination of the azimuth of a sound. When a sound is present on the contralateral side, most cells are excited, while ipsilateral sounds cause inhibition. It has been shown that the cells can go from complete excitation to full inhibition with a difference of only a few hundred milliseconds in arrival time of the two inputs.

The question then arises: How does the nervous system put together the signals available to it so that a determination of output can take place? To arrive at an understanding of how the nervous system intergrades incoming information at a given moment of time, we must understand that the processes that take place depend both on cellular forms and a topologic architecture and on the physiologic properties that relate input to output. That is, we have to know the *transfer* functions or *coupling* functions. Integration depends on the weighting of inputs. One of the important factors determining weighting is the area of synaptic contact. The extensive dendrites are the primary integrating structures. Electronic spread is the means of mixing, smoothing, attenuating, delaying, and summing postsynaptic potentials. The spatial distribution of input is often not random but systematically restricted. Also, the wide variety of characteristic geometries of synapses is no doubt important not only for the weighting of different combinations of inputs. When repeated stimuli are presented at various intervals at different junctions, increasing synaptic potentials are generated if the intervals between them are not too short or too long. This increase is due to a phenomenon called *facilitation*. If the response lasts longer than the interval between impulses, such that the second response rises from the residue of the first, then it is temporal summation. If, in addition, the response increment due to the second stimulus is larger than the preceding one, then it is facilitation. Facilitation is an important function of the nervous system and is found in quite different forms and durations ranging from a few milliseconds to tenths of seconds. Facilitation may grade from forms of sensitization to learning, especially at long intervals. A special case is the so-called *posttetanic potentiation* that is the result of high-frequency stimulation for long periods of time (about 10 seconds). This is an interesting case, since no effects can be seen during stimulation, but afterwards, any test stimulus at various intervals creates a marked increase in response up to many times more than the "tetanic" stimulus. *Antifacilitation* is the phenomenon where a decrease of response from the neuron is observed at certain junctions due to successive impulses. Its mechanism is less understood than facilitation. Both facilitation and antifacilitation may be observed on the same neuron but in different functions of it.

3.3   Representation of Information in the Nervous System

Whenever information is transferred between different parts of the nervous system, some communication paths have to be established, and some parameters of impulse firing relevant to communication must be set up. Since what is communicated is nothing more than impulses — spike trains — the only basic variables in a train of events are the number and intervals between spikes. With respect to this, the nervous system acts like a pulse-coded analog device, since the intervals are continuously graded. There exists a distribution of interval lengths between individual spikes, which in any sample can be expressed by the shape of the interval histogram. If one examines different examples, their distributions differ markedly. Some histograms look like Poisson distributions; some others exhibit gaussian or bimodal shapes. The coefficient of variation — expressed as the standard deviation over the mean — in some cases is constant, while in others it varies. Some other properties depend on the sequence of longer and shorter intervals than the mean. Some neurons show no linear dependence; some others show positive or negative correlations of successive intervals. If a stimulus is delivered and a discharge from the neuron is observed, a *poststimulus time histogram* can be used, employing the onset of the stimulus as a reference point and averaging many responses in order to reveal certain consistent features of temporal patterns. Coding of information can then be based on the average frequency, which can represent relevant gradations of the input. Mean frequency is the code in most cases, although no definition of it has been given with respect to measured quantities, such as averaging time, weighting functions, and forgetting functions. Characteristic transfer functions have been found, which suggests that there are several distinct coding principles in addition to the mean frequency. Each theoretically possible code becomes a candidate code as long as there exists some evidence that is readable by the system under investigation. Therefore, one has to first test for the availability of the code by imposing a stimulus that is considered "normal." After a response has been observed, the code is considered to be available. If the input is then changed to different levels of one parameter and changes are observed at the postsynaptic level, the code is called *readable.* However, only if both are formed in the same preparation and no other parameter is available and readable can the code be said to be the *actual* code employed. Some such parameters are

1. Time of firing
2. Temporal pattern
3. Number of spikes in the train
4. Variance of interspike intervals
5. Spike delays or latencies
6. Constellation code

The latter is a very important parameter, especially when used in conjunction with the concept of *receptive fields* of units in the different sensory pathways. The unit receptors do not need to have highly specialized abilities to permit encoding of a large number of distinct stimuli. Receptive fields are topographic and overlap extensively. Any given stimulus will excite a certain constellation of receptors and is therefore encoded in the particular set that is activated. A large degree of uncertainty prevails and requires the brain to operate probabilistically. In the nervous system there exists a large amount of *redundancy,* although neurons might have different thresholds. It is questionable, however, if these units are entirely equivalent, although they share parts of their receptive fields. The nonoverlapping parts might be of importance and critical to sensory function. On the other hand, redundancy does not necessarily mean unspecified or random connectivity. Rather, it allows for greater sensitivity and resolution, improvement of signal-to-noise ratio, while at the same time it provides stability of performance.

Integration of large numbers of converging inputs to give a single output can be considered as an averaging or probabilistic operation. The "decisions" made by a unit depend on its inputs, or some intrinsic states, and reaching a certain threshold. This way every unit in the nervous system can make a decision when it changes from one state to a different one. A theoretical possibility also exists that a mass of randomly connected neurons may constitute a trigger unit and that activity with a sharp threshold can spread through such a mass redundancy. Each part of the nervous system, and in particular the receiving side, can be thought of as a filter. Higher-order neurons do not merely pass their information on, but instead they use convergence from different channels, as well as divergence of the same channels and other processes, in order to modify incoming signals. Depending on the structure and coupling functions of the network, what gets through is determined. Similar networks exist at the output side. They also act as filters, but since they formulate decisions and commands with precise *spatiotemporal* properties, they can be thought of as *pattern generators.*

3.4   Lateral Inhibition

This discussion would be incomplete without a description of a very important phenomenon in the nervous system. This phenomenon, called *lateral inhibition,* used by the nervous system to improve spatial resolution and contrast. The effectiveness of this type of inhibition decreases with distance. In the retina, for example, lateral inhibition is used extensively in order to improve contrast. As the stimulus approaches a certain unit, it first excites neighbors of the recorded cell. Since these neighbors inhibit that unit, it responds by a decrease in firing frequency. If the stimulus is exactly over the recorded unit, this unit is excited and fires above its normal rate, and as the stimulus moves out again, the neighbors are excited, while the unit under consideration fires less. If we now examine the output of all the units as a whole and at one while half the considered array is stimulated and the other half is not, we will notice that at the point of discontinuity of the stimulus going from stimulation to nonstimulation, the firing frequencies of the two halves have been differentiated to the extreme at the stimulus edge, which has been enhanced. The neuronal circuits responsible for lateral shifts are relatively simple. Lateral inhibition can be considered to give the negative of the second spatial derivative of the input stimulus. A second layer of neurons could be constructed to perform this spacial differentiation on the input signal to detect the edge only. It is probably lateral inhibition that explains the psychophysical illusion known as *Mach bands.* It is probably the same principle that operates widely in the nervous system to enhance the sensitivity to contrast in the visual system in particular and in all other modalities in general. Through the years, different models have been developed to describe lateral inhibition mathematically, and various methods of analysis have been employed. Such methods include

Functional notations
Graphic solutions
Tabular solution
Taylor's series expansions
Artificial neural network modeling

These models include both one-dimensional examination of the phenomenon and two-dimensional treatment, where a two-dimensional array is used as a stimulus. This two-dimensional treatment is justified because most of the sensory receptors of the body form two-dimensional maps (receptive fields). In principle, if a one-dimensional lateral inhibition system is linear, one can extend the analysis to two dimensions by means of superposition. The two-dimensional array can be thought of as a function $f(x, y)$, and the lateral inhibition network itself is embodied in a separate $N \times N$ array, the central square of which has a positive value and can be thought of as a direct input from an incoming axon. The surrounding squares have negative values that are higher than the corner values, which are also negative. The method consists of multiplying the input signal values $f(x, y)$ and their contiguous values by the lateral inhibitory network's weighting factors to get a corresponding $g(x, y)$. Figure 3.1 presents an example of such a process. The technique illustrated here is used in the contrast enhancement of photographs. The objective is the same as that of the nervous system: to improve image sharpness without introducing too much distortion. This technique requires storage of each picture element and lateral "inhibitory" interactions between adjacent elements. Since a picture may contain millions of elements, high-speed computers with large-scale memories are required.

At a higher level, similar algorithms can be used to evaluate decision-making mechanisms. In this case, many inputs from different sensory systems are competing for attention. The brain evaluates each one of the inputs as a function of the remaining ones. One can picture a decision-making mechanism resembling a "locator" of stimulus peaks. The final output depends on what weights are used at the inputs of a push-pull mechanism. Thus a decision can be made depending on the weights an individual's brain

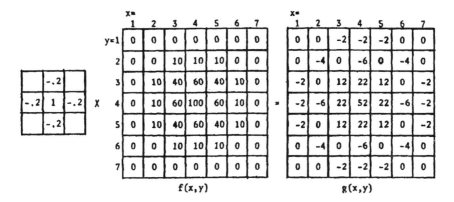

FIGURE 3.1 An example of two-dimensional lateral inhibition. On the left, the 3 × 3 array corresponds to the values of the synaptic junctions weighting coefficients. For simplicity, the corner weights are assumed to be zero. $g(x, y)$ represents the output matrix after lateral inhibition has been applied to the input matrix.

is applying to the incoming information about a situation under consideration. The most important information is heavily weighted, while the rest is either totally masked or weighted very lightly.

3.5 Higher Functions of the Nervous System

Pattern Recognition

One way of understanding human perception is to study the mechanism of information processing in the brain. The recognition of patterns of sensory input is one of the functions of the brain, a task accomplished by neuronal circuits, the *feature extractors*. Although such neuronal information is more likely to be processed globally by a large number of neurons, in animals, single-unit recording is one of the most powerful tools in the hands of the physiologist. Most often, the concept of the *receptive field* is used as a method of understanding sensory information processing. In the case of the visual system, one could call the receptive field a well-defined region of the visual field which, when stimulated, will change the firing rate of a neuron in the visual pathway. The response of that neuron will usually depend on the distribution of light in the receptive field. Therefore, the information collected by the brain from the outside world is transformed into spatial as well as temporal patterns of neuronal activity.

The question often asked is how do we perceive and recognize faces, objects, and scenes. Even in those cases where only noisy representations exist, we are still able to make some inference as to what the pattern represents. Unfortunately, in humans, single-unit recording, as mentioned above, is impossible. As a result, one has to use other kinds of measurements, such as *evoked potentials (EPs)*. Although physiologic in nature, EPs are still far away from giving us information at the neuronal level. Yet EPs have been used extensively as a way of probing the human (and animal) brains because of their noninvasive character. EPs can be considered to be the result of integrations of the neuronal activity of many neurons some place in the brain. This gross potential can then be used as a measure of the response of the brain to sensory input.

The question then becomes: Can we somehow use this response to influence the brain in producing patterns of activity that we want? None of the efforts of the past closed this loop. How do we explain then the phenomenon of selective attention by which we selectively direct our attention to something of interest and discard the rest? And what happens with the evolution of certain species that change appearance according to their everyday needs? All these questions tend to lead to the fact that somewhere in the brain there is a loop where previous knowledge or experience is used as a feedback to the brain itself. This feedback then modifies the ability of the brain to respond in a different way to the same

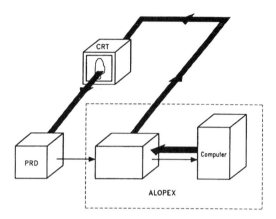

FIGURE 3.2   An ALOPEX system. The stimulus is presented on the CRT. The observer or any pattern-recognition device (PRD) faces the CRT; the subject's response is sent to the ALOPEX interface unit, where it is recorded and integrated, and the final response is sent to the computer. The computer calculates the values of the new pattern to be presented on the CRT according to the ALOPEX algorithm, and the process continues until the desired pattern appears on the CRT. At this point, the response is considered to be optimal and the process stops.

stimulus the next time it is presented. In a way, then, the brain creates mental "images" (independent of the stimulus) that tend to modify the representation of the stimulus in the brain.

This section describes some efforts in which different methods have been used in trying to address the difficult task of feedback loops in the brain. However, no attempt will be made to explain or even postulate where these feedback loops might be located. If one considers the brain as a huge set of neural nets, then one question has been debated for many years: What is the role of the individual neuron in the net, and what is the role of each network in the holistic process of the brain? More specifically, does the neuron act as an analyzer or a detector of specific features, or does it merely reflect the characteristic response of a population of cells of which it happens to be a member? What invariant relationships exist between sensory input and the response of a single neuron, and how much can be "read" about the stimulus parameters from the record of a single EP? In turn, then, how much feedback can one use from a single EP in order to influence the stimulus, and how successful can that influence be? Many physiologists express doubts that simultaneous observations of large numbers of individual neuronal activities can be readily interpreted. The main question we are asking is: Can a feedback process influence and modulate the stimuli patterns so that they appear optimal? If this is proven to be true, it would mean that we can reverse the pattern-recognition process, and instead of recognizing a pattern, we would be able to create a pattern from a vast variety of possible patterns. It would be like creating a link between our brain and a computer; equivalent to a brain-computer system network. Figure 3.2 is a schematic representation of such a process involved in what we call the *feedback loop* of the system. The pattern-recognition device (PRD) is connected to an ALOPEX system (a computer algorithm and an image processor in this case) and faces a display monitor where different intensity patterns can be shown. Thin arrows representing response information and heavy arrows representing detailed pattern information are generated by the computer and relayed by the ALOPEX system to the monitor. ALOPEX is a set of algorithms described in detail elsewhere in this handbook. If this kind of arrangement is used for the determination of visual receptive fields of neurons, then the PRD is nothing more than the brain of an experimental animal. This way the neuron under investigation does its own selection of the best stimulus or trigger feature and reverses the role of the neuron from being a feature extractor to becoming a feature generator, as mentioned earlier. The idea is to find the response of the neuron to a stimulus and use this response as a positive feedback in the directed evaluation of the initially random pattern. Thus the cell involved filters out the key trigger features from the stimulus and reinforces them with the feedback.

As a generalization of this process, one might consider that a neuron N receives a visual input from a pattern P, which is transmitted in a modified form P', to an analyzer neuron AN (or even a complex

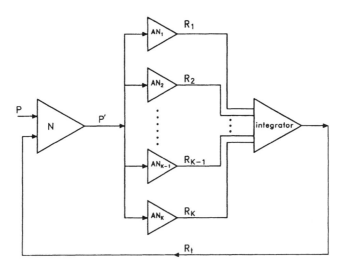

FIGURE 3.3 Diagammatic representation of the ALOPEX "inverse" pattern-recognition scheme. Each neuron represents a feature analyzer that responds to the stimulus with a scalar quantity R called the *response*. R is then fed back to the system, and the pattern is modified accordingly. This process continues until there is a close correlation between the desired output and the original pattern.

of neurons), as shown in Fig. 3.3. The analyzer responds with a scalar variable R that is then fed back to the system, and the pattern is modified accordingly. The process continues in small steps until there is an almost perfect correlation between the original pattern (template) and the one that neuron N indirectly created. This integrator sends the response back to the original modifier. The integrator need not be a linear summator. It could take any nonlinear form, a fact that is a more realistic representation of the visual cortex. One can envision the input patterns as templates preexisting in the memory of the system, a situation that might come about with visual experience. For a "naive" system, any initial pattern will do. As experience is gained, the patterns become less random. If one starts with a pattern that has some resemblance to one of the preexisting patterns, evolution will take its course. In nature, there might exist a mechanism similar to that of ALOPEX. By filtering the characteristics most important for the survival of the species, changes would be triggered. Perception, therefore, could be considered to be an interaction between sensory inputs and past experience in the form of templates stored in the memory of the perceiver and specific to the perceiver's needs. These templates are modifiable with time and adjusted accordingly to the input stimuli. With this approach, the neural nets and ensembles of nets under observation generate patterns that describe their thinking and memory properties. The normal flow of information is reversed and controls the afferent systems.

The perception processes as well as feature extraction or suppression of images or objects can be ascribed to specific neural mechanisms due to some sensory input or even due to some "wishful thinking" of the PRD. If it is true that the association cortex is affecting the sensitivity of the sensory cortex, then an ALOPEX mechanism is what one needs to close the loop for memory and learning.

Memory and Learning

If we try to define what memory is, we will face the fact that memory is not a single mental faculty but is rather composed of multiple abilities mediated by separate and distinct brain systems. Memory for a recent event can be expressed *explicitly* as a conscious recollection or *implicitly* as a facilitation of test performance without conscious recollection. The major distinction between these two memories is that explicit or *declarative* memory depends on limbic and diencephalic structures and provides the basis for recollection of events, while implicit or *nondeclarative* memory supports skills and habit learning, single conditioning, and the well-researched phenomenon of *priming*.

Declarative memory refers to memory of recent events and is usually assessed by tests of recall or recognition for specific single items. When the list of items becomes longer, a subject not only learns about each item on the list but also makes associations about what all these items have in common; i.e., the subject learns about the category that the items belong to. Learning leads to changes that increase or decrease the effectiveness of impulses arriving at the junctions between neurons, and the cumulative effect of these changes constitutes memory. Very often a particular pattern of neural activity leads to a result that occurs some time after the activity has ended. Learning then requires some means of relating the activity that is to be changed to the evaluation that can be made only by the delayed consequence. This phenomenon in physics is called *hysteresis* and refers to any modifications of future actions due to past actions. *Learning* then could be defined as change in any neuronal response resulting from previous experiences due to an external stimulus. Memory, in turn, would be the maintenance of these changes over time. The collection of neural changes representing memory is commonly known as the *engram,* and a major part of recent work has been to identify and locate engrams in the brain, since specific parts of the nervous system are capable of specific types of learning. The view of memory that has recently emerged is that information storage is tied to specific processing areas that are engaged during learning. The brain is organized so that separate regions of neocortex simultaneously carry out computations on specific features or characteristics of the external stimulus, no matter how complex that stimulus might be. If the brain learns specific properties or features of the stimulus, then we talk about the *nonassociative memory.* Associated with this type of learning is the phenomenon of *habituation,* in which if the same stimulus is presented repeatedly, the neurons respond less and less, while the introduction of a new stimulus increases the sensitization of the neuron. If the learning includes two related stimuli, then we talk about associative learning. This type of learning includes two types: *classic conditioning* and *operant conditioning.* The first deals with relationships among stimuli, while the latter deals with the relationship of the stimulus to the animal's own behavior. In humans, there exist two types of memory: *short-term* and *long-term memories.* The best way to study any physiologic process in humans, and especially memory, is to study its pathology. The study of amnesia has provided strong evidence distinguishing between these types of memory. Amnesic patients can keep a short list of numbers in mind for several minutes if they pay attention to the task. The difficulty comes when the list becomes longer, especially if the amount to be learned exceeds the brain capacity of what can be held in immediate memory. It could be that this happens because more systems have to be involved and that temporary information storage may occur within each brain area where stable changes in synaptic efficacy can eventually develop. *Plasticity* within existing pathways can account for most of the observations, and short-term memory occurs too quickly for it to require any major modifications of neuronal pathways. The capacity of long-term memory requires the integrity of the medial temporal and diencephalic regions in conjunction with neurons for storage of information. Within the domain of long-term memory, amnesic patients demonstrate intact learning and retention of certain motor, perceptual, and cognitive skills and intact priming effects. These patients do not exhibit any learning deficits but have no conscious awareness of prior study sessions or recognition of previously presented stimuli.

Priming effects can be tested by presenting words and then providing either the first few letters of the word or the last part of the word for recognition by the patient. Normal subjects, as expected, perform better than amnesic subjects. However, if these patients are instructed to "read" the incomplete word instead of memorizing it, then they perform as well as the normal individuals. Also, these amnesic patients perform well if words are cued by category names. Thus priming effects seem to be independent of the processes of recall and recognition memory, which is also observed in normal subjects. All this evidence supports the notion that the brain has organized its memory functions around fundamentally different information storage systems. In perceiving a word, a preexisting array of neurons is activated that have concurrent activities that produce perception, and priming is one of these functions.

Memory is not fixed immediately after learning but continues to grow toward stabilization over a period of time. This stabilization is called *consolidation of memory.* Memory consolidation is a *dynamic* feature of long-term memory, especially the declarative memory, but it is neither an automatic process with fixed lifetime nor is it determined at the time of learning. It is rather a process of reorganization of stored information. As time passes, some not yet consolidated memories fade out by remodeling the neural circuitry that is responsible for the original representation or by establishing new representations, since the original one can be forgotten.

The problems of learning and memory are studied continuously and with increased interest these days, especially because artificial systems such as neural networks can be used to mimic functions of the nervous system.

References

Cowan WM, Cuenod M (Eds). 1975. *Use of Axonal Transport for Studies of Neuronal Connectivity.* New York, Elsevier.

Deutsch S, Micheli-Tzanakou E. 1987. *Neuroelectric Systems.* New York, NYU Press.

Ganong WF. 1989. *Review of Medical Physiology,* 14th ed. Norwalk, CT, Appleton and Lange.

Hartzell HC. 1981. Mechanisms of slow postsynaptic potentials. *Nature* 291:593.

McMahon TA. 1984. *Muscles, Reflexes and Locomotion.* Princeton, NJ, Princeton University Press.

Partridge LD, Partridge DL. 1993. *The Nervous System: Its Function and Interaction with the World.* Cambridge, MA, MIT Press.

Shepherd GM. 1978. Microcircuits in the nervous system. *Sci Am* 238(2):92–103.

4

Vision System

George Stetten
Duke University

David Marr, an early pioneer in computer vision, defined *vision* as extracting "… from images of the external world, a description that is useful for the viewer and not cluttered with irrelevant information" [Marr, 1982]. Advances in computers and video technology in the past decades have created the expectation that artificial vision should be realizable. The nontriviality of the task is evidenced by the continuing proliferation of new and different approaches to computer vision without any observable application in our everyday lives. Actually, computer vision is already offering practical solutions in industrial assembly and inspection, as well as for military and medical applications, so it seems we are beginning to master some of the fundamentals. However, we have a long way to go to match the vision capabilities of a 4-year-old child. In this chapter we will explore what is known about how nature has succeeded at this formidable task — that of interpreting the visual world.

4.1 Fundamentals of Vision Research

Research into biologic vision systems has followed several distinct approaches. The oldest is psychophysics, in which human and animal subjects are presented with visual stimuli and their responses recorded. Important early insights also were garnered by correlating clinical observations of visual defects with known neuroanatomic injury. In the past 50 years, a more detailed approach to understanding the mechanisms of vision has been undertaken by inserting small electrodes deep within the living brain to monitor the electrical activity of individual neurons and by using dyes and biochemical markers to track the anatomic course of nerve tracts. This research has led to a detailed and coherent, if not complete, theory of a visual system capable of explaining the discrimination of form, color, motion, and depth. This theory has been confirmed by noninvasive radiologic techniques that have been used recently to study the physiologic responses of the visual system, including positron emission tomography [Zeki et al., 1991] and functional magnetic resonance imaging [Belliveau et al., 1992; Cohen and Bookheimer, 1994], although these noninvasive techniques provide far less spatial resolution and thus can only show general regions of activity in the brain.

4.2 A Modular View of the Vision System

The Eyes

Movement of the eyes is essential to vision, not only allowing rapid location and tracking of objects but also preventing stationary images on the retina, which are essentially invisible. Continual movement of the image on the retina is essential to the visual system.

The eyeball is spherical and therefore free to turn in both horizontal and vertical directions. Each eye is rotated by three pairs of mutually opposing muscles, innervated by the oculomotor nuclei in the brainstem. The eyes are coordinated as a pair in two useful ways: turning together to find and follow objects and turning inward to allow adjustment for parallax as objects become closer. The latter is called *convergence.*

The optical portion of the eye, which puts an image on the retina, is closely analogous to a photographic or television camera. Light enters the eye, passing through a series of transparent layers — the cornea, the aqueous humor, the lens, and the vitreous body — to eventually project on the retina.

The *cornea*, the protective outer layer of the eye, is heavily innervated with sensory neurons, triggering the blink reflex and tear duct secretion in response to irritation. The cornea is also an essential optical element, supplying two thirds of the total refraction in the eye. Behind the cornea is a clear fluid, the *aqueous humor*, in which the central aperture of the iris, the pupil, is free to constrict or dilate. The two actions are accomplished by opposing sets of muscles.

The *lens*, a flexible transparent object behind the iris, provides the remainder of refraction necessary to focus an image on the retina. The ciliary muscles surrounding the lens can increase the lens' curvature, thereby decreasing its focal length and bringing nearer objects into focus. This is called *accommodation*. When the ciliary muscles are at rest, distant objects are in focus. There are no contradictory muscles to flatten the lens. This depends simply on the elasticity of the lens, which decreases with age. Behind the lens is the *vitreous humor*, consisting of a semigelatinous material filling the volume between the lens and the retina.

The Retina

The retina coats the back of the eye and is therefore spherical, not flat, making optical magnification constant at 3.5 degrees of scan angle per millimeter. The retina is the neuronal front end of the visual system, the image sensor. In addition, it accomplishes the first steps in edge detection and color analysis before sending the processed information along the optic nerve to the brain. The retina contains five major classes of cells, roughly organized into layers. The dendrites of these cells each occupy no more than 1 to 2 mm^2 in the retina, limiting the extent of spatial integration from one layer of the retina to the next.

First come the *receptors*, which number approximately 125 million in each eye and contain the light-sensitive pigments responsible for converting photons into chemical energy. Receptor cells are of two general varieties: *rods* and *cones*. The cones are responsible for the perception of color, and they function only in bright light. When the light is dim, only rods are sensitive enough to respond. Exposure to a single photon may result in a measurable increase in the membrane potential of a rod. This sensitivity is the result of a chemical cascade, similar in operation to the photo multiplier tube, in which a single photon generates a cascade of electrons. All rods use the same pigment, whereas three different pigments are found in three separate kinds of cones.

Examination of the retina with an otoscope reveals its gross topography. The yellow circular area occupying the central 5 degrees of the retina is called the *macula lutea*, within which a small circular pit called the *fovea* may be seen. Detailed vision occurs only in the fovea, where a dense concentration of cones provides visual activity to the central 1 degree of the visual field.

On the inner layer of the retina is a layer of *ganglion cells*, whose axons make up the optic nerve, the output of the retina. They number approximately 1 million, or less than 1% of the number of receptor cells. Clearly, some data compression has occurred in the space between the receptors and the ganglion cells. Traversing this space are the *bipolar cells*, which run from the receptors through the retina to the ganglion cells. Bipolar cells exhibit the first level of information processing in the visual system; namely, their response to light on the retina demonstrates "center/surround" receptive fields; that is, a small dot on the retina elicits a response, while the area surrounding the spot elicits the opposite response. If both the center and the surround are illuminated, the net result is no response. Thus bipolar cells respond only at the border between dark and light areas. Bipolar cells come in two varieties, on-center and off-center, with the center respectively brighter or darker than the surround.

The center response of bipolar cells results from direct contact with the receptors. The surround response is supplied by the *horizontal cells,* which run parallel to the surface of the retina between the receptor layer and the bipolar layer, allowing the surrounding area to oppose the influence of the center. The *amacrine cells,* a final cell type, also run parallel to the surface but in a different layer, between the bipolar cells and the ganglion cells, and are possibly involved in the detection of motion.

Ganglion cells, since they are triggered by bipolar cells, also have center/surround receptive fields and come in two types, on-center and off-center. On-center ganglion cells have a receptive field in which illumination of the center increases the firing rate and a surround where it decreases the rate. Off-center ganglion cells display the opposite behavior. Both types of ganglion cells produce little or no change in firing rate when the entire receptive field is illuminated, because the center and surround cancel each other. As in many other areas of the nervous system, the fibers of the optic nerve use frequency encoding to represent a scalar quantity.

Multiple ganglion cells may receive output from the same receptor, since many receptive fields overlap. However, this does not limit overall spatial resolution, which is maximum in the fovea, where two points separated by 0.5 minutes of arc may be discriminated. This separation corresponds to a distance on the retina of 2.5 μm, which is approximately the center-to-center spacing between cones. Spatial resolution falls off as one moves away from the fovea into the peripheral vision, where resolution is as low as 1 degree of arc.

Several aspects of this natural design deserve consideration. Why do we have center/surround receptive fields? The ganglion cells, whose axons make up the optic nerve, do not fire unless there is meaningful information, i.e., a border, falling within the receptive field. It is the edge of a shape we see rather than its interior. This represents a form of data compression. Center/surround receptive fields also allow for relative rather than absolute measurements of color and brightness. This is essential for analyzing the image independent of lighting conditions. Why do we have both on-center and off-center cells? Evidently, both light and dark are considered information. The same shape is detected whether it is lighter or darker than the background.

Optic Chiasm

The two optic nerves, from the left and right eyes, join at the optic chiasm, forming a *hemidecussation,* meaning that half the axons cross while the rest proceed uncrossed. The resulting two bundles of axons leaving the chiasm are called the *optic tracts.* The left optic tract contains only axons from the left half of each retina. Since the images are reversed by the lens, this represents light from the right side of the visual field. The division between the right and left optic tracts splits the retina down the middle, bisecting the fovea. The segregation of sensory information into the contralateral hemispheres corresponds to the general organization of sensory and motor centers in the brain.

Each optic tract has two major destinations on its side of the brain: (1) the superior colliculus and (2) the lateral geniculate nucleus (LGN). Although topographic mapping from the retina is scrambled within the optic tract, it is reestablished in both major destinations so that right, left, up, and down in the image correspond to specific directions within those anatomic structures.

Superior Colliculus

The *superior colliculus* is a small pair of bumps on the dorsal surface of the midbrain. Another pair, the *inferior colliculus,* is found just below it. Stimulation of the superior colliculus results in contralateral eye movement. Anatomically, output tracts from the superior colliculus run to areas that control eye and neck movement. Both the inferior and superior colliculi are apparently involved in locating sound. In the bat, the inferior colliculus is enormous, crucial to that animal's remarkable echolocation abilities. The superior colliculus processes information from the inferior colliculus, as well as from the retina, allowing the eyes to quickly find and follow targets based on visual and auditory cues.

Different types of eye movements have been classified. The *saccade* (French, for "jolt") is a quick motion of the eyes over a significant distance. The saccade is how the eyes explore an image, jumping from landmark to landmark, rarely stopping in featureless areas. *Nystagmus* is the smooth pursuit of a moving image, usually with periodic backward saccades to lock onto subsequent points as the image moves by. *Microsaccades* are small movements, several times per second, over 1 to 2 minutes of arc in a seemingly random direction. Microsaccades are necessary for sight; their stabilization leads to effective blindness.

LGN

The thalamus is often called "the gateway to the cortex" because it processes much of the sensory information reaching the brain. Within the thalamus, we find the *lateral geniculate nucleus* (LGN), a peanut-sized structure that contains a single synaptic stage in the major pathway of visual information to higher centers. The LGN also receives information back from the cortex, so-called reentrant connections, as well as from the nuclei in the brainstem that control attention and arousal.

The cells in the LGN are organized into three pairs of layers. Each pair contains two layers, one from each eye. The upper two pairs consist of parvocellular cells (*P cells*) that respond with preference to different colors. The remaining lower pair consists of magnocellular cells (*M cells*) with no color preference (Fig. 4.1). The topographic mapping is identical for all six layers; i.e., passing through the layers at a given point yields synapses responding to a single area of the retina. Axons from the LGN proceed to the primary visual cortex in broad bands, the *optic radiations,* preserving this topographic mapping and displaying the same center/surround response as the ganglion cells.

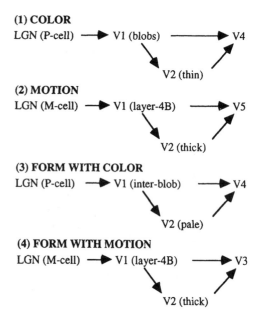

FIGURE 4.1   Visual pathways to cortical areas showing the separation of information by type. The lateral geniculate nucleus (LGN) and areas V1 and V2 act as gateways to more specialized higher areas.

Area V1

The LGN contains approximately 1.5 million cells. By comparison, the *primary visual cortex,* or *striate cortex,* which receives the visual information from the LGN, contains 200 million cells. It consists of a thin (2-mm) layer of gray matter (neuronal cell bodies) over a thicker collection of white matter (myelinated axons) and occupies a few square inches of the occipital lobes. The primary visual cortex has been called *area 17* from the days when the cortical areas were first differentiated by their cytoarchitectonics (the microscopic architecture of their layered neurons). In modern terminology, the primary visual cortex is often called *visual area 1,* or simply *V1.*

Destroying any small piece of V1 eliminates a small area in the visual field, resulting in *scotoma,* a local blind spot. Clinical evidence has long been available that a scotoma may result from injury, stroke, or tumor in a local part of V1. Between neighboring cells in V1's gray matter, horizontal connections are at most 2 to 5 mm in length. Thus, at any given time, the image from the retina is analyzed piecemeal in V1. Topographic mapping from the retina is preserved in great detail. Such mapping is seen elsewhere in the brain, such as in the somatosensory cortex [Mountcastle, 1957]. Like all cortical surfaces, V1 is a highly convoluted sheet, with much of its area hidden within its folds. If unfolded, V1 would be roughly pear-shaped, with the top of the pear processing information from the fovea and the bottom of the pear processing the peripheral vision. Circling the pear at a given latitude would correspond roughly to circling the fovea at a fixed radius.

The primary visual cortex contains six layers, numbered 1 through 6. Distinct functional and anatomic types of cells are found in each layer. Layer 4 contains neurons that receive information from the LGN. Beyond the initial synapses, cells demonstrate progressively more complex responses. The outputs of V1 project to an area known as *visual area 2* (*V2*), which surrounds V1, and to higher visual areas in the occipital, temporal, and parietal lobes as well as to the superior colliculus. V1 also sends reentrant projections back to the LGN. Reentrant projections are present at almost every level of the visual system [Felleman and Essen, 1991; Edelman, 1978].

Cells in V1 have been studied extensively in animals by inserting small electrodes into the living brain (with surprisingly little damage) and monitoring the individual responses of neurons to visual stimuli. Various subpopulations of cortical cells have thus been identified. Some, termed *simple cells*, respond to illuminated edges or bars at specific locations and at specific angular orientations in the visual field. The angular orientation must be correct within 10 to 20 degrees for the particular cell to respond. All orientations are equally represented. Moving the electrode parallel to the surface yields a smooth rotation in the orientation of cell responses by about 10 degrees for each 50 μm that the electrode is advanced. This rotation is subject to reversals in direction, as well as "fractures," or sudden jumps in orientation.

Other cells, more common than simple cells, are termed *complex cells*. Complex cells respond to a set of closely spaced parallel edges within a particular receptive field. They may respond specifically to movement perpendicular to the orientation of the edge. Some prefer one direction of movement to the other. Some complex and simple cells are *end-stopped*, meaning they fire only if the illuminated bar or edge does not extend too far. Presumably, these cells detect corners, curves, or discontinuities in borders and lines. End-stopping takes place in layers 2 and 3 of the primary visual cortex. From the LGN through the simple cells and complex cells, there appears to be a sequential processing of the image. It is probable that simple cells combine the responses of adjacent LGN cells and that complex cells combine the responses of adjacent simple cells.

A remarkable feature in the organization of V1 is binocular convergence, in which a single neuron responds to identical receptive fields in both eyes, including location, orientation, and directional sensitivity to motion. It does not occur in the LGN, where axons from the left and right eyes are still segregated into different layers. Surprisingly, binocular connections to neurons are present in V1 at birth. Some binocular neurons are equally weighted in terms of responsiveness to both eyes, while others are more sensitive to one eye than to the other. One finds columns containing the latter type of cells in which one eye dominates, called *ocular dominance columns*, in uniform bands approximately 0.5 mm wide everywhere in V1. Ocular dominance columns occur in adjacent pairs, one for each eye, and are prominent in animals with forward-facing eyes, such as cats, chimpanzees, and humans. They are nearly absent in rodents and other animals whose eyes face outward.

The topography of orientation-specific cells and of ocular dominance columns is remarkably uniform throughout V1, which is surprising because the receptive fields near the fovea are 10 to 30 times smaller than those at the periphery. This phenomenon is called magnification. The fovea maps to a greater relative distance on the surface of V1 than does the peripheral retina, by as much as 36-fold [Daniel and Whitteridge, 1961]. In fact, the majority of V1 processes only the central 10 degrees of the visual field. Both simple and complex cells in the foveal portion can resolve bars as narrow as 2 minutes of arc. Toward the periphery, the resolution falls off to 1 degree of arc.

As an electrode is passed down through the cortex *perpendicular* to the surface, each layer demonstrates receptive fields of characteristic size, the smallest being at layer 4, the input layer. Receptive fields are larger in other layers due to lateral integration of information. Passing the electrode *parallel* to the surface of the cortex reveals another important uniformity to V1. For example, in layer 3, which sends output fibers to higher cortical centers, one must move the electrode approximately 2 mm to pass from one collection of receptive fields to another that does not overlap. An area approximately 2 mm across thus represents the smallest unit piece of V1, i.e., that which can completely process the visual information. Indeed, it is just the right size to contain a complete set of orientations and more than enough to contain information from both eyes. It receives a few tens of thousands of fibers from the LGN, produces perhaps 50,000 output fibers, and is fairly constant in cytoarchitectonics whether at the center of vision, where it processes approximately 30 minutes of arc, or at the far periphery, where it processes 7 to 8 degrees of arc.

The topographic mapping of the visual field onto the cortex suffers an abrupt discontinuity between the left and right hemispheres, and yet our perception of the visual scene suffers no obvious rift in the midline. This is due to the *corpus callosum,* an enormous tract containing at least 200 million axons, that connects the two hemispheres. The posterior portion of the corpus callosum connects the two halves of V1, linking cells that have similar orientations and whose receptive fields overlap in the vertical midline. Thus a perceptually seamless merging of left and right visual fields is achieved. Higher levels of the visual system are likewise connected across the corpus callosum. This is demonstrated, for example, by the clinical observation that cutting the corpus callosum prevents a subject from verbally describing objects in the left field of view (the right hemisphere). Speech, which normally involves the left hemisphere, cannot process visual objects from the right hemisphere without the corpus callosum.

By merging the information from both eyes, V1 is capable of analyzing the distance to an object. Many cues for depth are available to the visual system, including occlusion, parallax (detected by the convergence of the eyes), optical focusing of the lens, rotation of objects, expected size of objects, shape based on perspective, and shadow casting. Stereopsis, which uses the slight difference between images due to the parallax between the two eyes, was first enunciated in 1838 by Sir Charles Wheatstone and is probably the most important cue [Wheatstone, 1838]. Fixating on an object causes it to fall on the two foveas. Other objects that are nearer become outwardly displaced on the two retinas, while objects that are farther away become inwardly displaced. About 2 degrees of horizontal disparity is tolerated, with fusion by the visual system into a single object. Greater horizontal disparity results in double vision. Almost no vertical displacement (a few minutes of arc) is tolerated. Physiologic experiments have revealed a particular class of complex cells in V1 that are *disparity tuned.* They fall into three general classes. One class fires only when the object is at the fixation distance, another only when the object is nearer, and a third only when it is farther away [Poggio and Talbot, 1981]. Severing the corpus callosum leads to a loss of stereopsis in the vertical midline of the visual field.

When the inputs to the two retinas cannot be combined, one or the other image is rejected. This phenomenon is known as *retinal rivalry* and can occur in a piecewise manner or can even lead to blindness in one eye. The general term *amblyopia* refers to the partial or complete loss of eyesight not caused by abnormalities in the eye. The most common form of amblyopia is caused by *strabismus,* in which the eyes are not aimed in a parallel direction but rather are turned inward (cross-eyed) or outward (wall-eyed). This condition leads to habitual suppression of vision from one of the eyes and sometimes to blindness in that eye or to *alternation,* in which the subject maintains vision in both eyes by using only one eye at a time. Cutting selected ocular muscles in kittens causes strabismus, and the kittens respond by alternation, preserving functional vision in both eyes. However, the number of cells in the cortex displaying binocular responses is greatly reduced. In humans with long-standing alternating strabismus, surgical repair making the eyes parallel again does not bring back a sense of depth. Permanent damage has been caused by the subtle condition of the images on the two retinas not coinciding. This may be explained by the Hebb model for associative learning, in which temporal association between inputs strengthens synaptic connections [Hebb, 1961].

Further evidence that successful development of the visual system depends on proper input comes from clinical experience with children who have *cataracts* at birth. Cataracts constitute a clouding of the lens, permitting light, but not images, to reach the retina. If surgery to remove the cataracts is delayed until the child is several years old, the child remains blind even though images are restored to the retina. Kittens and monkeys whose eyelids are sewn shut during a critical period of early development stay blind even when the eyes are opened. Physiologic studies in these animals show very few cells responding in the visual cortex. Other experiments depriving more specific elements of an image, such as certain orientations or motion in a certain direction, yield a cortex without the corresponding cell type.

Color

Cones, which dominate the fovea, can detect wavelengths between 400 and 700 nm. The population of cones in the retina can be divided into three categories, each containing a different pigment. This was established by direct microscopic illumination of the retina [Wald, 1974; Marks et al., 1964]. The pigments have a bandwidth on the order of 100 nm, with significant overlap, and with peak sensitivities at 560 nm (yellow-green), 530 nm (blue-green), and 430 nm (violet). These three cases are commonly known as red, green, and blue. Compared with the auditory system, whose array of cochlear sensors can discriminate thousands of different sonic frequencies, the visual system is relatively impoverished with only three frequency parameters. Instead, the retina expends most of its resolution on spatial information. Color vision is absent in many species, including cats, dogs, and some primates, as well as in most nocturnal animals, since cones are useless in low light.

By having three types of cones at a given locality on the retina, a simplified spectrum can be sensed and represented by three independent variables, a concept known as *trichromacy*. This model was developed by Thomas Young and Hermann von Helmholtz in the 19th century before neurobiology existed and does quite well at explaining the retina [Young, 1802; Helmholtz, 1889]. The model is also the underlying basis for red-green-blue (RGB) video monitors and color television [Ennes, 1981]. Rods do not help in discriminating color, even though the pigment in rods does add a fourth independent sensitivity peak.

Psychophysical experimentation yields a complex, redundant map between spectrum and perceived color, or *hue*, including not only the standard red, orange, yellow, green, and blue but hues such as pink, purple, brown, and olive green that are not themselves in the rainbow. Some of these may be achieved by introducing two more variables: *saturation*, which allows for mixing with white light, and *intensity*, which controls the level of color. Thus three variables are still involved: hue, saturation, and intensity.

Another model for color vision was put forth in the 19th century by Ewald Hering [Hering, 1864]. This theory also adheres to the concept of trichromacy, espousing three independent variables. However, unlike the Young-Helmholtz model, these variables are signed; they can be positive, negative, or zero. The resulting three axes are *red-green*, *yellow-blue*, and *black-white*. The Hering model is supported by the physiologic evidence for the center/surround response, which allows for positive as well as negative information. In fact, two populations of cells, activated and suppressed along the red-green and yellow-blue axes, have been found in monkey LGN. Yellow is apparently detected by a combination of red and green cones.

The Hering model explains, for example, the perception of the color brown, which results only when orange or yellow is surrounded by a brighter color. It also accounts for the phenomenon of color constancy, in which the perceived color of an object remains unchanged under differing ambient light conditions provided background colors are available for comparison. Research into color constancy was pioneered in the laboratory of Edwin Land [Land and McCann, 1971]. As David Hubel says, "We require color borders for color, just as we require luminance borders for black and white" [Hubel, 1988, p. 178]. As one might expect, when the corpus callosum is surgically severed, color constancy is absent across the midline.

Color processing in V1 is confined to small circular areas, known as *blobs*, in which *double-opponent cells* are found. They display a center/surround behavior based on the red-green and yellow-blue axes but lack orientation selectivity. The V1 blobs were first identified by their uptake of certain enzymes, and only later was their role in color vision discovered [Livingstone and Hubel, 1984]. The blobs are especially prominent in layers 2 and 3, which receive input from the P cells of the LGN.

Higher Cortical Centers

How are the primitive elements of image processing so far discussed united into an understanding of the image? Beyond V1 are many higher cortical centers for visual processing, at least 12 in the occipital lobe and others in the temporal and parietal lobes. Area V2 receives axons from both the blob and interblob areas of V1 and performs analytic functions such as filling in the missing segments of an edge. V2 contains three areas categorized by different kinds of stripes: *thick stripes* that process relative horizontal position and stereopsis, *thin stripes* that process color without orientations, and *pale stripes* that extend the process of end-stopped orientation cells.

Beyond V2, higher centers have been labeled V3, V4, V5, etc. Four parallel systems have been delineated [Zeki, 1992], each system responsible for a different attribute of vision, as shown in Fig. 4.1. This is obviously an oversimplification of a tremendously complex system.

Corroborative clinical evidence supports this model. For example, lesions in V4 lead to *achromatopsia*, in which a patient can only see gray and cannot even recall colors. Conversely, a form of poisoning, *carbon monoxide chromatopsia*, results when the V1 blobs and V2 thin stripes selectively survive exposure to carbon monoxide thanks to their rich vasculature, leaving the patient with a sense of color but not of shape. A lesion in V5 leads to *akinetopsia*, in which objects disappear.

As depicted in Fig. 4.1, all visual information is processed through V1 and V2, although discrete channels within these areas keep different types of information separate. A total lesion of V1 results in the perception of total blindness. However, not all channels are shown in Fig. 4.1, and such a "totally blind" patient may perform better than randomly when forced to guess between colors or between motion in different directions. The patient with this condition, called *blindsight,* will deny being able to see anything [Weiskrantz, 1990].

Area V1 preserves retinal topographic mapping and shows receptive fields, suggesting a piecewise analysis of the image, although a given area of V1 receives sequential information from disparate areas of the visual environment as the eyes move. V2 and higher visual centers show progressively larger receptive fields and less defined topographic mapping but more specialized responses. In the extreme of specialization, neurobiologists joke about the "grandmother cell," which would respond only to a particular face. No such cell has yet been found. However, cortical regions that respond to faces in general have been found in the temporal lobe. Rather than a "grandmother cell," it seems that face-selective neurons are members of ensembles for coding facts [Gross and Sergen, 1992].

Defining Terms

Binocular convergence: The response of a single neuron to the same location in the visual field of each eye.

Color constancy: The perception that the color of an object remains constant under different lighting conditions. Even though the spectrum reaching the eye from that object can be vastly different, other objects in the field of view are used to compare.

Cytoarchitectonics: The organization of neuron types into layers as seen by various staining techniques under the microscope. Electrophysiologic responses of individual cells can be correlated with their individual layer.

Magnification: The variation in amount of retinal area represented per unit area of V1 from the fovea to the peripheral vision. Even though the fovea takes up an inordinate percentage of V1 compared with the rest of the visual field, the scale of the cellular organization remains constant. Thus the image from the fovea is, in effect, magnified before processing.

Receptive field: The area in the visual field that evokes a response in a neuron. Receptive fields may respond to specific stimuli such as illuminated bars or edges with particular directions of motion, etc.

Stereopsis: The determination of distance to objects based on relative displacement on the two retinas because of parallax.

Topographic mapping: The one-to-one correspondence between location on the retina and location within a structure in the brain. Topographic mapping further implies that contiguous areas on the retina map to contiguous areas in the particular brain structure.

References

Belliveau JH, Kwong KK et al. 1992. Magnetic resonance imaging mapping of brain function: Human visual cortex. *Invest Radiol* 27(suppl 2):S59.

Cohen MS, Bookheimer SY. 1994. Localization of brain function using magnetic resonance imaging. *Trends Neurosci* 17(7):268.

Daniel PM, Whitteridge D. 1961. The representation of the visual field on the cerebral cortex in monkeys. *J Physiol* 159:203.

Edelman GM. 1978. Group selection and phasic reentrant signalling: A theory of higher brain function. In GM Edelman and VB Mountcastle (eds), *The Mindful Brain*, pp 51–100, Cambridge, MIT Press.

Ennes HE. 1981. NTSC color fundamentals. In *Television Broadcasting: Equipment, Systems, and Operating Fundamentals*. Indianapolis, Howard W. Sams & Co.

Felleman DJ, V Essen DC. 1991. Distributed hierarchical processing in the primate cerebral cortex. *Cerebral Cortex* 1(1):1.

Gross CG, Sergen J. 1992. Face recognition. *Curr Opin Neurobiol* 2(2):156.

Hebb DO. 1961. *The Organization of Behavior*. New York, Wiley.

Helmholtz H. 1889, *Popular Scientific Lectures*. London, Longmans.

Hering E. 1864. *Outlines of a Theory of Light Sense*. Cambridge, Harvard University Press.

Hubel DH. 1995. *Eye, Brain, and Vision*. New York, Scientific American Library.

Land EH, McCann JJ. 1971. Lightness and retinex theory. *J. Opt Soc Am* 61:1.

Livingstone MS, Hubel DH. 1984. Anatomy and physiology of a color system in the primate visual cortex. *J Neurosci* 4:309.

Marks WB, Dobelle WH, MacNichol EF. 1964. Visual pigments of single primate cones. *Science* 143:1181.

Marr D. 1982. *Vision*. San Francisco, WH Freeman.

Mountcastle VB. 1957. Modality and topographic properties of single neurons of cat's somatic sensory cortex. *J Neurophysiol* 20(3):408.

Poggio GF, Talbot WH. 1981. Mechanisms of static and dynamic stereopsis in foveal cortex of the rhesus monkey. *J Physiol* 315:469.

Wald G. 1974. Proceedings: Visual pigments and photoreceptors — Review and outlook. *Exp Eye Res* 18(3):333.

Weiskrantz L. 1990. The Ferrier Lecture: Outlooks for blindsight: explicit methodologies for implicit processors. *Proc R Soc Lond* B239:247.

Wheatstone SC. 1838. Contribution to the physiology of vision. *Philosoph Trans R Soc Lond*

Young T. 1802. The Bakerian Lecture: On the theory of lights and colours. *Philosoph Trans R Soc Lond* 92:12.

Zeki S. 1992. The visual image in mind and brain. *Sci Am*, Sept. 1992, p. 69.

Zeki S, Watson JD, Lueck CJ, et al. 1991. A direct demonstration of functional specialization in human visual cortex. *J Neurosci* 11(3):641.

Further Reading

An excellent introductory text about the visual system is *Eye, Brain, and Vision,* by Nobel laureate, David H. Hubel (1995, Scientific American Library, New York). A more recent general text with a thorough treatment of color vision, as well as the higher cortical centers, is *A Vision of the Brain,* by Semir Zeki (1993, Blackwell Scientific Publications, Oxford).

Other useful texts with greater detail about the nervous system are *From Neuron to Brain,* by Nicholls, Martin, Wallace, and Kuffler (3rd ed., 1992, Sinauer Assoc., Sunderland, Mass.), *The Synaptic Organization of the Brain,* by Shepherd (4th ed., 1998, Oxford Press, New York), and *Fundamental Neuroanatomy,* by Nauta and Feirtag (1986, Freeman, New York).

A classic text that laid the foundation of computer vision by *Vision,* by David Marr (1982, Freeman, New York). Other texts dealing with the mathematics of image processing and image analysis are *Digital Image Processing,* by Pratt (1991, Wiley, New York), and *Digital Imaging Processing and Computer Vision,* by Schalkoff (1989, Wiley, New York).

5

Auditory System

Ben M. Clopton
University of Washington

Francis A. Spelman
University of Washington

The auditory system can be divided into two large subsystems, peripheral and central. The peripheral auditory system converts the condensations and rarefactions that produce sound into neural codes that are interpreted by the central auditory system as specific sound tokens that may affect behavior.

The peripheral auditory system is subdivided into the external ear, the middle ear, and the inner ear (Fig. 5.1). The external ear collects sound energy as pressure waves that are converted to mechanical motion at the *eardrum*. This motion is transformed across the *middle ear* and transferred to the *inner ear,* where it is frequency-analyzed and converted into neural codes that are carried by the eighth cranial nerve, or *auditory nerve,* to the central auditory system.

Sound information, encoded as discharges in an array of thousands of auditory nerve fibers, is processed in nuclei that make up the central auditory system. The major centers include the *cochlear nuclei* (CN), the *superior olivary complex* (SOC), the *nuclei of the lateral lemniscus* (NLL), the *inferior colliculi* (IC), the *medial geniculate body* (MGB) of the thalamus, and the *auditory cortex* (AC). The CN, SOC, and NLL are brainstem nuclei; the IC is at the midbrain level; and the MGB and AC constitute the auditory thalamocortical system.

While interesting data have been collected from groups other than mammals, this chapter will emphasize the mammalian auditory system. This chapter ignores the structure and function of the vestibular system. While a few specific references are included, most are general in order to provide a more introductory entry into topics.

5.1 Physical and Psychological Variables

Acoustics

Sound is produced by time-varying motion of the particles in air. The motions can be defined by their pressure variations or by their volume velocities. *Volume velocity* is defined as the average particle velocity produced across a cross-sectional area and is the acoustic analog of electric current. *Pressure* is the acoustic analog of voltage. *Acoustic intensity* is the average rate of the flow of energy through a unit area normal to the direction of the propagation of the sound wave. It is the product of the acoustic pressure and the

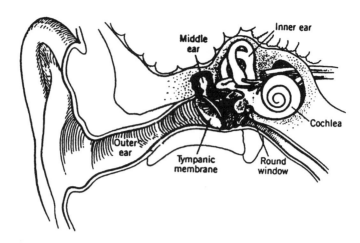

FIGURE 5.1 The peripheral auditory system showing the ear canal, tympanic membrane, middle ear and ossicles, and the inner ear consisting of the cochlea and semicircular canals of the vestibular system. Nerves communicating with the brain are also shown.

volume velocity and is analogous to electric power. *Acoustic impedance,* the analog of electrical impedance, is the complex ratio of acoustic pressure and volume velocity. Sound is often described in terms of either acoustic pressure or acoustic intensity [Kinsler and Frey, 1962].

The auditory system has a wide dynamic range, i.e., it responds to several decades of change in the magnitude of sound pressure. Because of this wide dynamic range, it is useful to describe the independent variables in terms of decibels, where acoustic intensity is described by $dB = 10 \log(I/I_0)$, where I_0 is the reference intensity, or equivalently for acoustic pressure, $dB = 20 \log(P/P_0)$, where P_0 is the reference pressure.

Psychoacoustics

Physical variables, such as *frequency* and *intensity,* may have correlated psychological variables, such as *pitch* and *loudness.* Relationships between acoustic and psychological variables, the subject of the field of *psychoacoustics,* are generally not linear and may be very complex, but measurements of human detection and discrimination can be made reliably. Humans without hearing loss detect tonal frequencies from 20 Hz to 20 kHz. At 2 to 4 kHz their *dynamic range,* the span between threshold and pain, is approximately 120 dB. The minimum threshold for sound occurs between 2 and 5 kHz and is about 20 µPa. At the low end of the auditory spectrum, threshold is 80 dB higher, while at the high end, it is 70 dB higher. Intensity differences of 1 dB can be detected, while frequency differences of 2 to 3 Hz can be detected at frequencies below about 3 kHz [Fay, 1988].

5.2   The Peripheral Auditory System

The External Ear

Ambient sounds are collected by the *pinna,* the visible portion of the external ear, and guided to the middle ear by the *external auditory meatus,* or ear canal. The pinna acquires sounds selectively due to its geometry and the sound shadowing effect produced by the head. In those species whose ears can be moved voluntarily through large angles, selective scanning of the auditory environment is possible.

The ear canal serves as an acoustic waveguide that is open at one end and closed at the other. The open end at the pinna approximates a short circuit (large volume velocity and small pressure variation), while that at the closed end is terminated by the *tympanic membrane* (eardrum). The tympanic membrane has a relatively high acoustic impedance compared with the characteristic impedance of the meatus and looks like an open circuit. Thus the ear canal can resonate at those frequencies for which its length is an odd number of quarter wavelengths. The first such frequency is at about 3 kHz in the human. The meatus is antiresonant for those frequencies for which its length is an integer number of half wavelengths. For a discussion of resonance and antiresonance in an acoustic waveguide, see a text on basic acoustics, e.g., Kinsler and Frey [1962].

The acoustic properties of the external ear produce differences between the sound pressure produced at the tympanic membrane and that at the opening of the ear canal. These differences are functions of frequency, with larger differences found at frequencies between 2 and 6 kHz than those below 2 kHz. These variations have an effect on the frequency selectivity of the overall auditory system.

The Middle Ear

Anatomy

Tracing the acoustic signal, the boundaries of the middle ear include the tympanic membrane at the input and the oval window at the output. The middle ear bones, the ossicles, lie between. Pressure relief for the tympanic membrane is provided by the eustachian tube. The middle ear is an air-filled cavity.

The Ossicles

The three bones that transfer sound from the tympanic membrane to the *oval window* are called the *malleus* (hammer), *incus* (anvil), and *stapes* (stirrup). The acoustic impedance of the atmospheric source is much less than that of the aqueous medium of the load. The ratio is 3700 in an open medium, or 36 dB [Kinsler and Frey, 1962]. The ossicles comprise an impedance transformer for sound, producing a mechanical advantage that allows the acoustic signal at the tympanic membrane to be transferred with low loss to the round window of the cochlea (inner ear). The air-based sound source produces an acoustic signal of low-pressure and high-volume velocity, while the mechanical properties of the inner ear demand a signal of high-pressure and low-volume velocity.

The impedance transformation is produced in two ways: The area of the tympanic membrane is greater than that of the footplate of the stapes, and the lengths of the malleus and incus produce a lever whose length is greater on the side of the tympanic membrane than it is on the side of the oval window. In the human, the mechanical advantage is about 22:1 [Dobie and Rubel, 1989] and the impedance ratio of the transformer is 480, 27 dB, changing the mismatch from 3700:1 to about 8:1.

This simplified discussion of the function of the ossicles holds at low frequencies, those below 2 kHz. First, the tympanic membrane does not behave as a piston at higher frequencies but can support modes of vibration. Second, the mass of the ossicles becomes significant. Third, the connections between the ossicles is not lossless, nor can the stiffness of these connections be ignored. Fourth, pressure variations in the middle ear cavity can change the stiffness of the tympanic membrane. Fifth, the cavity of the middle ear produces resonances at acoustic frequencies.

Pressure Relief

The eustachian tube is a bony channel that is lined with soft tissue. It extends from the middle ear to the nasopharynx and provides a means by which pressure can be equalized across the tympanic membrane. The function is clearly observed with changes in altitude or barometric pressure. A second function of the eustachian tube is to aerate the tissues of the middle ear.

The Inner Ear

The mammalian inner ear is a spiral structure, the *cochlea* (snail), consisting of three fluid-filled chambers, or scalae, the *scala vestibuli,* the *scala media,* and the *scala tympani* (Fig. 5.2). The stapes footplate introduces mechanical displacements into the scala vestibuli through the oval window at the *base* of the cochlea. At the other end of the spiral, the *apex* of the cochlea, the scala vestibuli and the scala tympani

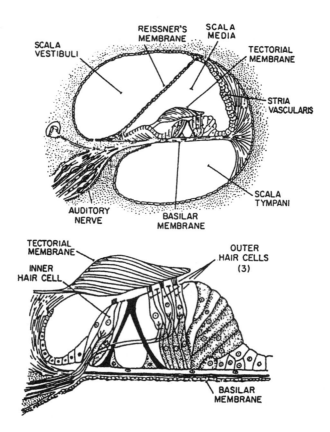

FIGURE 5.2 Cross-section of one turn of the cochlea showing the scala vestibuli, scala media, and scala tympani. Reissner's membrane separates the SM and SV, while the basilar membrane and organ of Corti separate the SM and ST.

communicate by an opening, the *helicotrema.* Both are filled with an aqueous medium, the *perilymph.* The scala media spirals between them and is filled with *endolymph,* a medium that is high in K^+ and low in Na^+. The scala media is separated from the scala vestibuli by *Reissner's membrane,* which is impermeable to ions, and the scala media is separated from the scala tympani by the *basilar membrane* (BM) and *organ of Corti.* The organ of Corti contains the hair cells that transduce acoustic signals into neural signals, the cells that support the hair cells, and the tectorial membrane to which the outer hair cells are attached. The BM provides the primary filter function of the inner ear and is permeable so that the cell bodies of the hair cells are bathed in perilymph.

Fluid and tissue displacements travel from the footplate of the stapes along the cochlear spiral from base to apex. Pressure relief is provided for the incompressible fluids of the inner ear by the round window membrane; e.g., if a transient pressure increase at the stapes displaces its footplate inward, there will be a compensatory outward displacement of the round window membrane.

The Basilar Membrane

Physiology

The BM supports the hair cells and their supporting cells (see Fig. 5.2). Sound decomposition into its frequency components is a major code of the BM. A transient sound, such as a click, initiates a *traveling wave* of displacement in the BM, and this motion has frequency-dependent characteristics that arise from properties of the membrane and its surrounding structures [Bekesy, 1960]. The membrane's width varies as it traverses the cochlear duct: It is narrower at its basal end than at its apical end. It is stiffer at the base than at the apex, with stiffness varying by about two orders of magnitude [Dobie and Rubel, 1989].

The membrane is a distributed structure, which acts as a delay line, as suggested by the nature of the traveling wave [Lyon and Mead, 1989]. The combination of mechanical properties of the BM produces a structure that demonstrates a distance-dependent displacement when the ear is excited sinusoidally. The distance from the apex to the maximum displacement is logarithmically related to the frequency of a sinusoidal tone [LePage, 1991].

Tuning is quite sharp for sinusoidal signals. The slope of the tuning curve is much greater at the high-frequency edge than at the low-frequency edge, with slopes of more than 100 dB per octave at the high edge and about half that at the low edge [Lyon and Mead, 1989]. The filter is sharp, with a 10-dB bandwidth of 10 to 25% of the center frequency.

The auditory system includes both passive and active properties. The outer hair cells (see below) receive efferent output from the brain and actively modify the characteristics of the auditory system. The result is to produce a "cochlear amplifier," which sharpens the tuning of the BM [Lyon and Mead, 1989], as well as adding nonlinear properties to the system [Geisler, 1992; Cooper and Rhode, 1992], along with otoacoustic emissions [LePage, 1991].

The Organ of Corti

The organ of Corti is attached to the BM on the side of the aqueous fluid of the scala media. It is comprised of the supporting cells for the hair cells, the hair cells themselves, and the *tectorial membrane*. The cilia of the *inner hair cells* (IHCs) do not contact the tectorial membrane, while those of the *outer hair cells* (OHCs) do. Both IHCs and OHCs have precise patterns of stereocilia at one end, which are held within the tectorial plate next to the overlying tectorial membrane. The IHCs synapse with *spiral ganglion cells,* the afferent neurons, while the OHCs synapse with efferent neurons. Both IHCs and OHCs are found along the length of the organ of Corti. The IHCs are found in a single line, numbering between about 3000 and 4000 in human. There are three lines of OHCs, numbering about 12,000 in human [Nadol, 1988].

Inner Hair Cells

The stereocilia of the IHCs are of graded, decreasing length from one side of the cell where a kinocilium is positioned early in ontogeny. If the cilia are deflected in a direction toward this position, membrane channels are further opened to allow potassium to enter and depolarize the cell [Hudspeth, 1987]. Displacement in the other direction reduces channel opening and produces a relative hyperpolarization [Hudspeth and Corey, 1977]. These changes in intracellular potential modulate transmitter release at the base of the IHCs.

The IHCs are not attached to the tectorial membrane, so their response to motion of the membrane is proportional to the velocity of displacement rather than to displacement itself, since the cilia of the hair cells are bathed in endolymph. When the membrane vibrates selectively in response to a pure tone, the stereocilia are bent atop a small number of hair cells, which depolarize in response to the mechanical event. Thus, the *tonotopic organization* of the BM is transferred to the hair cells and to the rest of the auditory system. The auditory system is organized tonotopically, i.e., in order of frequency, because the frequency ordering of the cochlea is mapped through successive levels of the system. While this organization is preserved throughout the system, it is much more complex than a huge set of finely tuned filters.

Hair cells in some species exhibit frequency tuning when isolated [Crawford and Fettiplace, 1985], but mammalian hair cells exhibit no tuning characteristics. The tuning of the mammalian auditory system depends on the mechanical characteristics of the BM as modified by the activity of the OHCs.

Outer Hair Cells

The OHCs have cilia that are attached to the tectorial membrane. Since their innervation is overwhelmingly efferent, they do not transfer information to the brain but are modulated in their mechanical action by the brain. There are several lines of evidence that lead to the conclusion that the OHCs play an active role in the processes of the inner ear. First, OHCs change their length in response to neurotransmitters

[Dobie and Rubel, 1989]. Second, observation of the Henson's cells, passive cells that are intimately connected to OHCs, shows that spontaneous vibrations are produced by the Henson's cells in mammals and likely in the OHCs as well. These vibrations exhibit spectral peaks that are appropriate in frequency to their locations on the BM [Khanna et al., 1993]. Third, action of the OHCs as amplifiers leads to spontaneous otoacoustic emissions [Kim, 1984; Lyon and Mead, 1989] and to changes in the response of the auditory system [Lyon and Mead, 1989; Geisler, 1992]. Fourth, AC excitation of the OHCs of mammals produces changes in length [Cooke, 1993].

The OHCs appear to affect the response of the auditory system in several ways. They enhance the tuning characteristics of the system to sinusoidal stimuli, decreasing thresholds and narrowing the filter's bandwidth [Dobie and Rubel, 1989]. They likely influence the damping of the BM dynamically by actively changing its stiffness.

Spiral Ganglion Cells and the Auditory Nerve

Anatomy

The auditory nerve of the human contains about 30,000 fibers consisting of myelinated proximal processes of spiral ganglion cells (SGCs). The somas of spiral ganglion cells (SGCs) lie in *Rosenthal's canal,* which spirals medial to the three scalae of the cochlea. Most (93%) are large, heavily myelinated, *type I* SGCs whose distal processes synapse on IHCs. The rest are smaller *type II* SGCs, which are more lightly myelinated. Each IHC has, on average, a number of fibers that synapse with it, 8 in the human and 18 in the cat, although some fibers contact more than one IHC. In contrast, each type II SGC contacts OHCs at a rate of about 10 to 60 cells per fiber.

The auditory nerve collects in the center of the cochlea, its *modiolus,* as SGC fibers join it. Low-frequency fibers from the apex join first, and successively higher frequency fibers come to lie concentrically on the outer layers of the nerve in a spiraling fashion before it exits the modiolus to enter the internal auditory meatus of the temporal bone. A precise tonotopic organization is retrained in the concentrically wrapped fibers.

Physiology

Discharge spike patterns from neurons can be recorded extracellularly while repeating tone bursts are presented. A *threshold level* can be identified from the resulting *rate-level function* (RLF). In the absence of sound and at lower, subthreshold levels, a *spontaneous rate* of discharge is measured. In the nerve this ranges from 50 spikes per second to less than 10. As intensity is raised, the *threshold level* is encountered, where the evoked discharge rate significantly exceeds the spontaneous discharge rate. The plot of threshold levels as a function of frequency is the neuron's *threshold tuning curve.* The tuning curves for axons in the auditory nerve show a minimal threshold (maximum sensitivity) at a *characteristic frequency* (CF) with a narrow frequency range of responding for slightly more intense sounds. At high intensities, a large range of frequencies elicits spike discharges. RLFs for nerve fibers are *monotonic* (i.e., spike rate increases with stimulus intensity), and although a saturation rate is usually approached at high levels, the spike rate does not decline. Mechanical tuning curves for the BM and neural threshold tuning curves are highly similar (Fig. 5.3). Mechanical frequency analysis in the cochlea and the orderly projection of fibers through the nerve lead to correspondingly orderly maps for CFs in the nerve and the nuclei of the central pathways.

Sachs and Young [1979] found that the frequency content of lower intensity vowel sounds is represented as corresponding tonotopic rate peaks in nerve activity, but for higher intensities this rate code is lost as fibers tend toward equal discharge rates. At high intensities spike synchrony to frequencies near CF continue to signal the relative spectral content of vowels, a temporal code. These results hold for *high-spontaneous-rate fibers* (over 15 spikes per second), which are numerous. Less common, *low-spontaneous-rate fibers* (less than 15 spikes per second) appear to maintain the rate code at higher intensities, suggesting different coding roles for these two fiber populations.

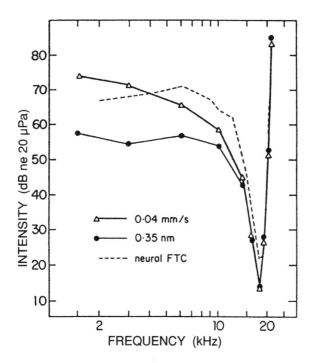

FIGURE 5.3   Mechanical and neural turning curves from the BM and auditory nerve, respectively. The two mechanical curves show the intensity and frequency combinations for tones required to obtain a criterion displacement or velocity, while the neural curve shows the combinations needed to increase neural discharge rates a small amount over spontaneous rate.

5.3   The Central Auditory System

Overview

In ascending paths, obligatory synapses occur at the CN, IC, MGB, and AC, but a large number of alternative paths exist with ascending and descending internuclear paths and the shorter intranuclear connections between neighboring neurons and subdivisions within a major nuclear group. Each of the centers listed contains subpopulations of neurons that differ in aspects of their morphologies, discharge patterns to sounds, segregation in the nucleus, biochemistry, and synaptic connectivities. The arrangement of the major ascending auditory pathways is schematically illustrated in Fig. 5.4. For references, see Altschuler et al. [1991].

Neural Bases of Processing

The Cochlear Nuclei

Anatomy of the Cochlear Nuclei.   The CN can be subdivided into at least three distinct regions, the *anteroventral CN* (AVCN), the *posteroventral CN* (PVCN), and the *dorsal CN* (DCN). Each subdivision has one or more distinctive neuron types and unique intra- and internuclear connections. The axon from each type I SGC in the nerve branches to reach each of the three divisions in an orderly manner so that tonotopic organization is maintained. Neurons with common morphologic classifications are found in all three divisions, especially *granule cells,* which tend to receive connections from type II spiral ganglion cells.

Morphologic classification of neurons based on the shapes of their dendritic trees and somas show that the anterior part of the AVCN contains many *spherical bushy cells,* while in its posterior part both

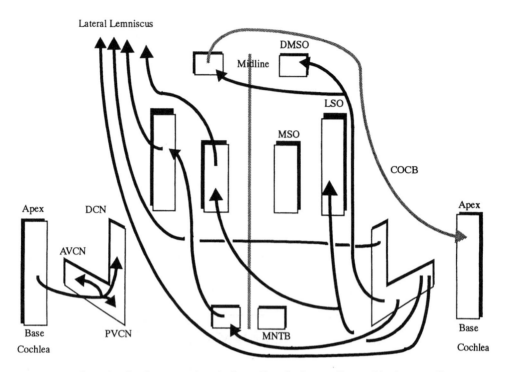

FIGURE 5.4   A schematic of major connections in the auditory brainstem discussed in the text. All structures and connections are bilaterally symmetrical, but connections have been shown on one side only for clarity. No cell types are indicated, but the subdivisions of origin are suggested in the CN. Note that the LSO and MSO receive inputs from both CN.

globular bushy cells and spherical bushy cells are found. Spherical bushy cells receive input from one type I ganglion cell through a large synapse formation containing end bulbs of Held, while the globular cells may receive inputs from a few afferent fibers. These endings cover a large part of the soma surface and parts of the proximal dendrite, especially in spherical bushy cells, and they have rounded vesicles presynaptically, indicating excitatory input to the bushy cells, while other synaptic endings of noncochlear origins tend to have flattened vesicles associated with inhibitory inputs. *Stellate cells* are found throughout the AVCN, as well as in the lower layers of the DCN. The AVCN is tonotopically organized, and neurons having similar CFs have been observed to lie in layers or laminae [Bourk et al., 1981]. *Isofrequency laminae* also have been indicated in other auditory nuclei.

The predominant neuron in the PVCN is the *octopus cell*, a label arising from its distinctive shape with asymmetrically placed dendrites. Octopus cells receive cochlear input from type I SGCs on their somas and proximal dendrites. Their dendrites cross the incoming array of cochlear fibers, and these often branch to follow the dendrite toward the soma.

The DCN is structurally the most intricate of the CN. In many species, four or five layers are noticeable, giving it a "cortical" structure, and its local circuitry has been compared with that of the cerebellum. *Fusiform cells* are the most common morphologic type. Their somas lie in the deeper layers of the DCN, and their dendrites extend toward the surface of the nucleus and receive primarily noncochlear inputs. Cochlear fibers pass deeply in the DCN and turn toward the surface to innervate fusiform and *giant cells* that lie in the deepest layer of the DCN. The axons of fusiform and giant cells project out of the DCN to the contralateral IC.

Intracellular recording in slice preparation is beginning to identify the membrane characteristics of neuronal types in the CN. The diversity of neuronal morphologic types, their participation in local circuits, and the emerging knowledge of their membrane biophysics are motivating detailed compartmental modeling [Arle and Kim, 1991].

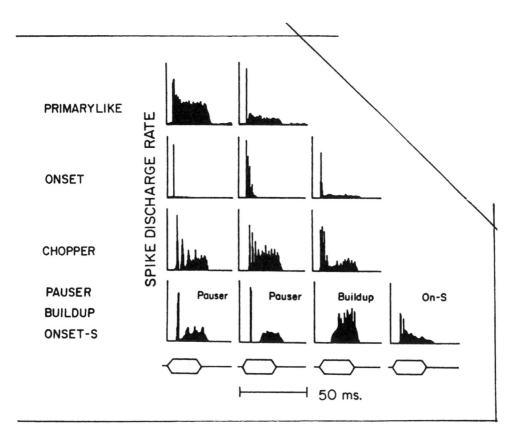

FIGURE 5.5 Peristimulus time histogram patterns obtained in the CN and nerve. Repeated presentations of a tone burst at CF are used to obtain these estimates of discharge rate during the stimulus. (Adapted from Young, 1984.)

Spike Discharge Patterns.   Auditory nerve fibers and neurons in central nuclei may discharge only a few times during a brief tone burst, but if a histogram of spike events is synchronized to the onset of the tone burst, a *peristimulus time histogram* (PSTH) is obtained that is more representative of the neuron's response than any single one. The PSTH may be expressed in terms of spike counts, spike probability, or spike rate as a function of time, but all these retain the underlying temporal pattern of the response. PSTHs taken at the CF for a tone burst intensity roughly 40 dB above threshold have shapes that are distinctive to different nuclear subdivisions and even different morphologic types. They have been used for functionally classifying auditory neurons.

Figure 5.5 illustrates some of the major pattern types obtained from the auditory nerve and CN. Auditory nerve fibers and spherical bushy cells in AVCN have *primary-like* patterns in their PSTHs, an elevated spike rate after tone onset, falling to a slowly adapting level until the tone burst ends. Globular bushy cells may have primary-like, *pri-notch* (primary-like with a brief notch after onset), or chopper patterns. Stellate cells have nonprimary-like patterns. *Onset* response patterns, one or a few brief peaks of discharge at onset with little or no discharges afterward, are observed in the PVCN from octopus cells. *Chopper, pauser,* and *buildup* patterns are observed in many cells of the DCN. For most neurons of the CN, these patterns are not necessarily stable over different stimulus intensities; a primary-like pattern may change to a pauser pattern and then to a chopper pattern as intensity is raised [Young, 1984].

Functional classification also has been based on the *response map,* a plot of a neuron's spike discharge rate as a function of tonal frequency and intensity. Fibers and neurons with primary-like PSTHs generally have response maps with only an *excitatory region* of elevated rate. The lower edges of this region approximate the threshold tuning curve. Octopus cells often have very broad tuning curves and extended response maps, as suggested by their frequency-spanning dendritic trees. More complex response maps are observed for some neurons, such as those in the DCN. Inhibitory regions alone, a frequency-intensity area of suppressed spontaneous discharge rates, or combinations of excitatory regions and inhibitory regions have been observed. Some neurons are excited only within islands of frequency-intensity combinations, demonstrating a CF but having no response to high-intensity sounds. In these cases, an RLF at CF would be *nonmonotonic;* i.e., spike rate decreases as the level is raised. Response maps in the DCN containing both excitatory and inhibitory regions have been shown to arise from a convergence of inputs from neurons with only excitatory or inhibitory regions in their maps [Young and Voigt, 1981].

Superior Olivary Complex (SOC)

The SOC contains 10 or more subdivisions in some species. It is the first site at which connections from the two ears converge and is therefore a center for binaural processing that underlies sound localization. There are large differences in the subdivisions between mammalian groups such as bats, primates, cetaceans, and burrowing rodents that utilize vastly different binaural cues. Binaural cues to the locus of sounds include *interaural level differences* (ILDs), *interaural time differences* (ITDs), and detailed spectral differences for multispectral sounds due to head and pinna filtering characteristics.

Neurons in the *medial superior olive* (MSO) and *lateral superior olive* (LSO) tend to process ITDs and ILDs, respectively. A neuron in the MSO receives projections from spherical bushy cells of the CN from both sides and thereby the precise timing and tuning cues of nerve fibers passed through the large synapses mentioned. The time accuracy of the pathways and the comparison precision of MSO neurons permit the discrimination of changes in ITD of a few tens of microseconds. MSO neurons project to the ipsilateral IC through the lateral lemniscus. Globular bushy cells of the CN project to the medial nucleus of the trapezoid body (MNTB) on the contralateral side, where they synapse on one and only one neuron in a large, excitatory synapse, the calyx of Held. MNTB neurons send inhibitory projections to neurons of the LSO on the same side, which also receives excitatory input from spherical bushy cells from the AVCN on the same side. Sounds reaching the ipsilateral side will excite discharges from an LSO neuron, while those reaching the contralateral side will inhibit its discharge. The relative balance of excitation and inhibition is a function of ILD over part of its physiological range, leading to this cue being encoded in discharge rate.

One of the subdivisions of the SOC, the *dorsomedial periolivary nucleus* (DMPO), is a source of efferent fibers that reach the contralateral cochlea in the *crossed olivocochlear bundle* (COCB). Neurons of the DMPO receive inputs from collaterals of globular bushy cell axons of the contralateral ACVN that project to the MNTB and from octopus cells on both sides. The functional role of the feedback from the DMPO to the cochlea is not well understood.

Nuclei of the Lateral Lemniscus (NLL)

The lateral lemniscus consists of ascending axons from the CN and LSO. The NLL lie within this tract, and some, such as the dorsal nucleus (DNLL), are known to process binaural information, but less is known about these nuclei as a group than others, partially due to their relative inaccessibility.

Inferior Colliculi (IC)

The IC are paired structures lying on the dorsal surface of the rostral brainstem. Each colliculus has a large *central nucleus* (ICC), a surface cortex, and paracentral nuclei. Each colliculus receives afferents from a number of lower brainstem nuclei, projects to the MGB through the *brachium,* and communicates with the other colliculus through a *commissure.* The ICC is the major division and has distinctive laminae in much of its volume. The laminae are formed from *disk-shaped cells* and afferent fibers. The disk-shaped cells, which make up about 80% of the ICC's neuronal population, have flattened dendritic fields that lie in the laminar plane. The terminal endings of afferents form fibrous layers between laminae. The remaining neurons in the ICC are *stellate cells* that have dendritic trees spanning laminae. Axons from these two cell types make up much of the ascending ICC output.

Tonotropic organization is preserved in the ICC's laminae, each corresponding to an *isofrequency lamina.* Both monaural and binaural information converges at the IC through direct projections from the CN and from the SOC and NLL. Crossed CN inputs and those from the ipsilateral MSO are excitatory. Inhibitory synapses in the ICC arise from the DNLL, mediated by gamma-amino-butyric acid (GABA), and from the ipsilateral LSO, mediated by glycine.

These connections provide an extensive base for identifying sound direction at this midbrain level, but due to their convergence, it is difficult to determine what binaural processing occurs at the IC as opposed to being passed from the SOC and NLL. Many neurons in the IC respond differently depending on binaural parameters. Varying ILDs for clicks or high-frequency tones often indicates that contralateral sound is excitatory. Ipsilateral sound may have no effect on responses to contralateral sound, classifying the cell as E0, or it may inhibit responses, in which case the neuron is classified as EI, or maximal excitation may occur for sound at both ears, classifying the neuron as EE. Neurons responding to lower frequencies are influenced by ITDs, specifically the phase difference between sinusoids at the ears. Spatial receptive fields for sounds are not well documented in the mammalian IC, but barn owls, who use the sounds of prey for hunting at night, have sound-based spatial maps in the homologous structure. The superior colliculus, situated just rostral to the IC, has spatial auditory receptive field maps for mammals and owl.

Auditory Thalamocortical System

Medial Geniculate Body (MGB).   The MGB and AC form the auditory thalamocortical system. As with other sensory systems, extensive projections to and from the cortical region exist in this system. The MGB has three divisions, the *ventral, dorsal,* and *medial.* The ventral division is the largest and has the most precise tonotopic organization. Almost all its input is from the ipsilateral ICC through the brachium of the IC. Its large *bushy cells* have dendrites oriented so as to lie in isofrequency layers, and the axons of these neurons project to the AC, terminating in layers III and IV.

Auditory Cortex.   The auditory cortex (AC) consists of areas of the cerebral cortex that respond to sounds. In mammals, the AC is bilateral and has a primary area with surrounding secondary areas. In nonprimates, the AC is on the lateral surface of the cortex, but in most primates, it lies within the lateral fissure on the superior surface of the temporal lobe. Figure 5.6 reveals the area of the temporal lobe involved with auditory function in humans. Tonotopic mapping is usually evident in these areas as isofrequency surface lines. The primary AC responds most strongly and quickly to sounds. In echo-locating bats, the cortex has a large tonotopic area devoted to the frequency region of its emitted cries and cues related to its frequency modulation and returned Doppler shift [Aitkin, 1990].

The cytoarchitecture of the primary AC shows layers I (surface) through VI (next to white matter), with the largest neurons in layers V and VI. Columns with widths of 50 to 75 μm are evident from dendritic organization in layers III and IV, with fibers lying between the columns. A description of cell types is beyond this treatment.

Discharge patterns in the AC for sound stimuli are mainly of the onset type. Continuous stimuli often evoke later spikes, after the onset discharge, in unanesthetized animals. About half the neurons in the primary AC have monotonic RLFs, but the proportion of nonmonotonic RLFs in secondary areas is much higher. A number of studies have used complex sounds to study cortical responses. Neural responses to species-specific cries, speech, and other important sounds have proven to be labile and to a great extent dependent on the arousal level and behavioral set of the animal.

Cortical lesions in humans rarely produce deafness, although deficits in speech comprehension and generation may exist. Audiometric tests will generally indicate that sensitivity to tonal stimuli is retained. It has been known for some time that left-hemisphere lesions in the temporal region can disrupt comprehension (Wernicke's area) and in the region anterior to the precentral gyrus (Broca's area) can interfere with speech production. It is difficult to separate the effects of these areas because speech comprehension provides vital feedback for producing correct speech.

5.4 Pathologies

Hearing loss results from conductive and neural deficits. Conductive hearing loss due to attenuation in the outer or middle ear often can be alleviated by amplification provided by hearing aids and may be

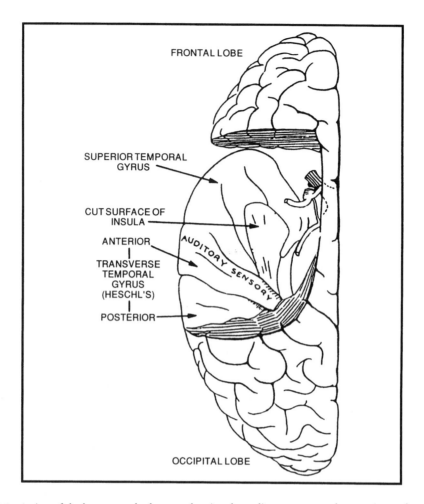

FIGURE 5.6 A view of the human cerebral cortex showing the auditory cortex on the superior surface of the left temporal lobe after removal of the overlying parietal cortex.

subject to surgical correction. Sensorineural loss due to the absence of IHCs results from genetic deficits, biochemical insult, exposure to intense sound, or aging (*presbycusis*). For some cases of sensorineural loss, partial hearing function can be restored with the cochlear prosthesis, electrical stimulation of remaining SGCs using small arrays of electrodes inserted into the scala tympani [Miller and Spelman, 1990]. In a few patients having no auditory nerve, direct electrical stimulation of the CN has been used experimentally to provide auditory sensation. Lesions of the nerve and central structures occur due to trauma, tumor growth, and vascular accidents. These may be subject to surgical intervention to prevent further damage and promote functional recovery.

5.5   Models of Auditory Function

Hearing mechanisms have been modeled for many years at a phenomenologic level using psychophysical data. As physiologic and anatomic observations have provided detailed parameters for peripheral and central processing, models of auditory encoding and processing have become more quantitative and physically based. Compartmental models of single neurons, especially SGCs and neurons in the CN, having accurate morphometric geometries, electrical properties, membrane biophysics, and local circuitry are seeing increasing use.

References

Aitkin L. 1990. *The Auditory Cortex: Structural and Functional Bases of Auditory Perception*. London, Chapman and Hall.

Altschuler RA, Bobbin RP, Clopton BM, Hoffman DW (Eds). 1991. *Neurobiology of Hearing: The Central Auditory System*. New York, Raven Press.

Arle JE, Kim DO. 1991. Neural modeling of intrinsic and spike-discharge properties of cochlear nucleus neurons. *Biol Cybern* 64:273.

Bekesy G. von. 1960. *Experiments in Hearing*. New York, McGraw-Hill.

Bourk TR, Mielcarz JP, Norris BE. 1981. Tonotopic organization of the anteroventral cochlear nucleus of the cat. *Hear Res* 4:215.

Cooke M. 1993. *Modelling Auditory Processing and Organisation*. Cambridge, England, Cambridge University Press.

Cooper NP, Rhode WS. 1992. Basilar membrane mechanics in the hook region of cat and guinea-pig cochleae: Sharp tuning and nonlinearity in the absence of baseline position shifts. *Hear Res* 63:163.

Crawford AC, Fettiplace R. 1985. The mechanical properties of ciliary bundles of turtle cochlear hair cells. *J Physiol* 364:359.

Dobie RA, Rubel EW. 1989. The auditory system: Acoustics, psychoacoustics, and the periphery. In HD Patton et al (Eds), *Textbook of Physiology, Vol 1: Excitable Cells and Neurophysiology, 21st ed.* Philadelphia, Saunders.

Fay RR. 1988. *Hearing in Vertebrates: A Psychophysics Databook*. Winnetka, Hill-Fay Associates.

Geisler CD. 1992. Two-tone suppression by a saturating feedback model of the cochlear partition. *Hear Res* 63:203.

Hudspeth AJ. 1987. Mechanoelectrical transduction by hair cells in the acousticolateralis sensory system. *Annu Rev Neurosci* 6:187.

Hudspeth AJ, Corey DP. 1977. Sensitivity, polarity, and conductance change in the response of vertebrate hair cells to controlled mechanical stimuli. *Proc Natl Acad Sci USA* 74:2407.

Khanna SM, Keilson SE, Ulfendahl M, Teich MC. 1993. Spontaneous cellular vibrations in the guinea-pig temporal-bone preparation. *Br J Audiol* 27:79.

Kim DO. 1984. Functional roles of the inner- and outer-hair-cell subsystems in the cochlea and brainstem. In CI Berlin (Ed), *Hearing Science: Recent Advances*. San Diego, Calif, College-Hill Press.

Kinsler LE, Frey AR. 1962. *Fundamentals of Acoustics*. New York, Wiley.

LePage EL. 1991. Helmholtz revisited: Direct mechanical data suggest a physical model for dynamic control of mapping frequency to place along the cochlear partition. In *Lecture Notes in Biomechanics*. New York, Springer-Verlag.

Lyon RF, Mead C. 1989. Electronic cochlea. In C Mead (Ed), *Analog VLSI and Neural Systems*. Reading, Mass. Addison-Wesley.

Miller JM, Spelman FA (Eds). 1990. *Cochlear Implants: Models of the Electrically Stimulated Ear*. New York, Springer-Verlag.

Nadol JB Jr. 1988. Comparative anatomy of the cochlea and auditory nerve in mammals. *Hear Res* 34:253.

Sachs MB, Young ED. 1979. Encoding of steady-state vowels in the auditory nerve: representation in terms of discharge rate. *J Acoust Soc Am* 66:470.

Young ED, Voigt HF. 1981. The internal organization of the dorsal cochlear nucleus. In J Syka and L Aitkin (Eds), *Neuronal Mechanisms in Hearing*, pp 127–133. New York, Plenum Press.

Young ED. 1984. Response characteristics of neurons of the cochlear nuclei. In CI Berlin (Ed), *Hearing Science: Recent Advances*. San Diego, Calif. College-Hill Press.

6

The Gastrointestinal System

Berj L. Bardakjian
University of Toronto

6.1 Introduction

The primary function of the gastrointestinal system (Fig. 6.1) is to supply the body with nutrients and water. The ingested food is moved along the alimentary canal at an appropriate rate for digestion, absorption, storage, and expulsion. To fulfill the various requirements of the system, each organ has adapted one or more functions. The esophagus acts as a conduit for the passage of food into the stomach for trituration and mixing. The ingested food is then emptied into the small intestine, which plays a major role in the digestion and absorption processes. The chyme is mixed thoroughly with secretions and it is propelled distally (1) to allow further gastric emptying, (2) to allow for uniform exposure to the absorptive mucosal surface of the small intestine, and (3) to empty into the colon. The vigor of mixing and the rate of propulsion depend on the required contact time of chyme with enzymes and the mucosal surface for efficient performance of digestion and absorption. The colon absorbs water and electrolytes from the chyme, concentrating and collecting waste products that are expelled from the system at appropriate times. All of these motor functions are performed by contractions of the muscle layers in the gastrointestinal wall.

6.2 Gastrointestinal Electrical Oscillations

Gastrointestinal motility is governed by myogenic, neural, and chemical control systems (Fig. 6.2). The myogenic control system is manifest by periodic depolarizations of the smooth muscle cells, which constitute autonomous electrical oscillations called the electrical control activity (ECA) or slow waves [Daniel and Chapman, 1963]. The properties of this myogenic system and its electrical oscillations dictate to a large extent the contraction patterns in the stomach, small intestine, and colon [Szurszewski, 1987]. The ECA controls the contractile excitability of smooth muscle cells since the cells may contract only

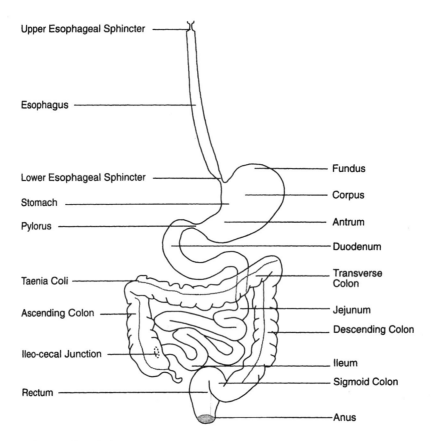

FIGURE 6.1   The gastrointestinal tract.

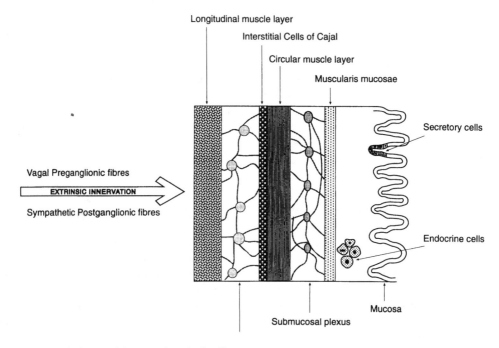

FIGURE 6.2   The layers of the gastrointestinal wall.

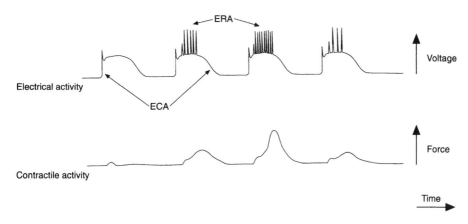

FIGURE 6.3 The relationships between ECA, ERA and muscular contractions. The ERA occurs in the depolarized phase of the ECA. Muscular contractions are associated with the ERA, and their amplitude depends on the frequency of response potentials within an ERA burst.

when depolarization of the membrane voltage exceeds an excitation threshold. The normal spontaneous amplitude of ECA depolarization does not exceed this excitation threshold except when neural or chemical excitation is present. The myogenic system affects the frequency, direction, and velocity of the contractions. It also affects the coordination or lack of coordination between adjacent segments of the gut wall. Hence, the electrical activities in the gut wall provide an electrical basis for gastrointestinal motility.

In the distal stomach, small intestine, and colon, there are intermittent bursts of rapid electrical oscillations, called the electrical response activity (ERA) or spike bursts. The ERA occurs during the depolarization plateaus of the ECA if a cholinergic stimulus is present, and it is associated with muscular contractions (Fig. 6.3). Thus, neural and chemical control systems determine whether contractions will occur or not, but when contractions are occurring, the myogenic control system (Fig. 6.4) determines the spatial and temporal patterns of contractions.

There is also a cyclical pattern of of distally propagating ERA that appears in the small intestine during the fasted state [Szurszewski, 1969], called the Migrating Motility Complex (MMC). This pattern consists of four phases [Code and Marlett, 1975]: Phase I has little or no ERA, phase II consists of irregular ERA bursts, phase III consists of intense repetitive ERA bursts where there is an ERA burst on each ECA cycle, and phase IV consists of irregular ERA bursts but is usually much shorter than phase II and may not be always present. The initiation and propagation of the MMC is controlled by enteric cholinergic neurons in the intestinal wall (Fig. 6.2). The propagation of the MMC may be modulated by inputs from extrinsic nerves or circulating hormones [Sarna et al., 1981]. The MMC keeps the small intestine clean of residual food, debris, and desquamated cells.

6.3 A Historical Perspective

Minute Rhythms

Alvarez and Mahoney [1922] reported the presence of a rhythmic electrical activity (which they called "action currents") in the smooth muscle layers of the stomach, small intestine, and colon. Their data was acquired from cat (stomach, small intestine), dog (stomach, small intestine, colon), and rabbit (small intestine, colon). They also demonstrated the existence of frequency gradients in excised stomach and bowel. Puestow [1933] confirmed the presence of a rhythmic electrical activity (which he called "waves of altered electrical potential") and a frequency gradient in isolated canine small intestinal segments. He also demonstrated the presence of an electrical spiking activity (associated with muscular contractions) superimposed on the rhythmic electrical activity. He implied that the rhythmic electrical activity persisted

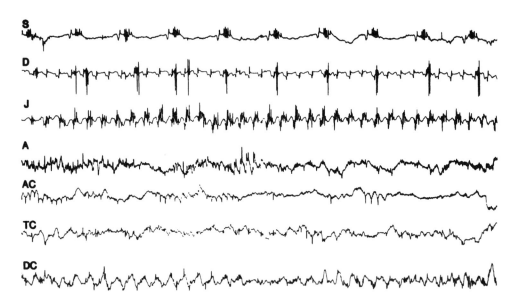

FIGURE 6.4 The gastrointestinal ECA and ERA, recorded in a conscious dog from electrode sets implanted sub-serosally on stomach (S), duodenum (D), jejunum (J), proximal ascending colon (A), distal ascending colon (AC), transverse colon (TC) and descending colon (DC), respectively. Each trace is of 2 min duration.

at all times, whereas the electrical spike activity was of an intermittent nature. Bozler [1938, 1939, 1941] confirmed the occurrence of an electrical spiking activity associated with muscular contractions both *in vitro* in isolated longitudinal muscle strips from guinea pig (colon, small intestine) and rabbit (small intestine), and *in situ* in exposed loops of small intestine of anesthetized cat, dog, and rabbit as well as in cat stomach. He also suggested that the strength of a spontaneous muscular contraction is proportional to the frequency and duration of the spikes associated with it.

The presence of two types of electrical activity in the smooth muscle layers of the gastrointestinal tract in several species had been established [Milton and Smith, 1956; Bulbring et al., 1958; Burnstock et al., 1963; Daniel and Chapman, 1963; Bass, 1965; Gillespie, 1962; Duthie, 1974; Christensen, 1975; Daniel, 1975; Sarna, 1975a]. The autonomous electrical rhythmic activity is an omnipresent myogenic activity [Burnstock et al., 1963] whose function is to control the appearance *in time and space* of the electrical spiking activity (an intermittent activity associated with muscular contractions) when neural and chemical factors are appropriate [Daniel and Chapman, 1963]. Neural and chemical factors determine whether or not contractions will occur, but when contractions are occurring, the myogenic control system determines the *spatial and temporal* patterns of contractions.

Isolation of a distal segment of canine small intestine from a proximal segment (using surgical transection or clamping) had been reported to produce a decrease in the frequency of both the rhythmic muscular contractions [Douglas, 1949; Milton and Smith, 1956] and the electrical rhythmic activity [Milton and Smith, 1956] of the distal segment, suggesting frequency entrainment or pulling of the distal segment by the proximal one. It was demonstrated [Milton and Smith, 1956] that the repetition of the electrical spiking activity changed in the same manner as that of the electrical rhythmic activity, thus confirming a one-to-one temporal relationship between the frequency of the electrical rhythmic activity, the repetition rate of the electrical spiking activity, and the frequency of the muscular contractions (when all are present at any one site). Nelson and Becker [1968] suggested that the electrical rhythmic activity of the small intestine behaves like a system of coupled relaxation oscillators. They used two forward coupled relaxation oscillators, having different intrinsic frequencies, to demonstrate frequency entrainment of the two coupled oscillators. Uncoupling of the two oscillators caused a decrease in the frequency of the distal oscillator simulating the effect of transection of the canine small intestine.

The electrical rhythmic activity in canine stomach [Sarna et al., 1972], canine small intestine [Nelson and Becker, 1968; Diamant et al., 1970; Sarna et al., 1971], human small intestine [Robertson-Dunn and Linkens, 1974], human colon [Bardakjian and Sarna, 1980], and human rectosigmoid [Linkens et al., 1976] has been modeled by populations of coupled nonlinear oscillators. The interaction between coupled nonlinear oscillators is governed by both intrinsic oscillator properties and coupling mechanisms.

Hour Rhythms

The existence of periodic gastric activity in the fasted state in both dog [Morat, 1882] and man [Morat, 1893] has been reported. The occurrence of a periodic pattern of motor activity, comprising bursts of contractions alternating with "intervals of repose," in the gastrointestinal tracts of fasted animals was noted early in the twentieth century by Boldireff [1905]. He observed that (1) the bursts recurred with a periodicity of about 1.5 to 2.5 h, (2) the amplitude of the gastric contractions during the bursts were larger than those seen postprandially, (3) the small bowel was also involved, and (4) with longer fasting periods, the bursts occurred less frequently and had a shorter duration. Periodic bursts of activity were also observed in (1) the lower esphageal sphincter [Cannon and Washburn, 1912], and (2) the pylorus [Wheelon and Thomas, 1921]. Further investigation of the fasting contractile activity in the upper small intestine was undertaken in the early 1920s with particular emphasis on the coordination between the stomach and duodenum [Wheelon and Thomas, 1922; Alvarez and Mahoney, 1923]. More recently, evidence was obtained [Itoh et al., 1978] that the cyclical activity in the lower esophageal sphincter noted by Cannon and Washburn [1912] was also coordinated with that of the stomach and small intestine.

With the use of implanted strain gauges, it was possible to observe contractile activity over long periods of time and it was demonstrated that the cyclical fasting pattern in the duodenum was altered by feeding [Jacoby et al., 1963]. The types of contractions observed during fasting and feeding were divided into four groups [Reinke et al., 1967; Carlson et al., 1972]. Three types of contractile patterns were observed in fasted animals: (1) quiescent interval, (2) a shorter interval of increasing activity, and (3) an interval of maximal activity. The fourth type was in fed animals and it consisted of randomly occurring contractions of varying amplitudes. With the use of implanted electrodes in the small intestine of fasted dogs, Szurszewski [1969] demonstrated that the cyclical appearance of electrical spiking activity at each electrode site was due to the migration of the cyclical pattern of quiescence, increasing activity and maximal electrical activity down the small intestine from the duodenum to the terminal ileum. He called this electrical pattern the *migrating myoelectric complex* (MMC). Grivel and Ruckebusch [1972] demonstrated that the mechanical correlate of this electrical pattern, which they called the *migrating motor complex*, occurs in other species such as sheep and rabbits. They also observed that the velocity of propagation of the maximal contractile activity was proportional to the length of the small intestine. Code and Marlett [1975] observed the electrical correlate of the cyclical activity in dog stomach that was reported by Morat [1882; 1893], and they demonstrated that the stomach MMC was coordinated with the duodenal MMC.

The MMC pattern has been demonstrated in other mammalian species [Ruckebusch and Fioramonti, 1975; Ruckebusch and Bueno, 1976], including humans. Bursts of distally propagating contractions has been noted in the gastrointestinal tract of man [Beck et al., 1965], and their cyclical nature was reported by Stanciu and Bennet [1975]. The MMC has been described in both normal volunteers [Vantrappen et al., 1977; Fleckenstein, 1978; Thompson et al., 1980; Kerlin and Phillips, 1982; Rees et al., 1982] and in patients [Vantrappen et al. 1977; Thompson et al., 1982; Summers et al., 1982].

Terminology

A nomenclature to describe the gastrointestinal electrical activities has been proposed to describe the minute rhythm [Sarna, 1975b] and the hour rhythm [Carlson et al., 1972; Code and Marlett, 1975].

Control cycle is one depolarization and repolarization of the transmembrane voltage. *Control wave (or slow wave)* is the continuing rhythmic electrical activity recorded at any one site. It was assumed to be generated by the smooth muscle cells behaving like a relaxation oscillator at that site. However, recent evidence [Hara et al., 1986; Suzuki et al. 1986; Barajas-Lopez et al., 1989; Serio et al., 1991] indicates that it is generated by a system of interstitial cells of Cajal (ICC) and smooth muscle cells at that site. *Electrical Control Activity (ECA)* is the totality of the control waves recorded at one or several sites. *Response Potentials (or spikes)* are the rapid oscillations of transmembrane voltage in the depolarized state of smooth muscle cells. They are associated with muscular contraction and their occurrence is assumed to be in response to a control cycle when acetylcholine is present. *Electrical Response Activity (ERA)* is the totality of the groups of response potentials at one or several sites.

Migrating Motility Complex (MMC) is the entire cycle, which is composed of four phases. Initially, the electrical and mechanical patterns were referred to as the migrating myoelectric complex and the migrating motor complex, respectively. *Phase I* is the interval during which fewer than 5% of ECA have associated ERA, and no or very few contractions are present. *Phase II* is the interval when 5 to 95% of the ECA has associated ERA, and intermittent contractions are present. *Phase III* is the interval when more than 95% of ECA have associated ERA, and large cyclical contractions are present. *Phase IV* is a short and waning interval of intermittent ERA and contractions. Phases II and IV are not always present and are difficult to characterize, whereas phases I and III are always present. *MMC Cycle Time* is the interval from the end of one phase III to the end of a subsequent phase III at any one site. *Migration Time* is the time taken for the MMC to migrate from the upper duodenum to the terminal ileum.

6.4 The Stomach

Anatomical Features

The stomach is somewhat pyriform in shape with its large end directed upward at the lower esophageal sphincter and its small end bent to the right at the pylorus. It has two curvatures, the greater curvature, which is four to five times as long as the lesser curvature, consists of three regions: the fundus, corpus (or body), and antrum, respectively. It has three smooth muscle layers. The outermost layer is the longitudinal muscle layer, the middle is the circular muscle layer and the innermost is the oblique muscle layer. These layers thicken gradually in the distal stomach towards the pylorus, which is consistent with stomach function since trituration occurs in the distal antrum. The size of the stomach varies considerably among subjects. In an adult male, its greatest length when distended is about 25 to 30 cm and its widest diameter is about 10 to 12 cm [Pick and Howden, 1977].

The structural relationships of nerve, muscle, and interstitial cells of Cajal in the canine corpus indicated a high density of gap junctions indicating very tight coupling between cells. Nerves in the corpus are not located close to circular muscle cells but are found exterior to the muscle bundles, whereas ICCs have gap junction contact with smooth muscle cells and are closely innervated [Daniel and Sakai, 1984].

Gastric ECA

In the canine stomach, the fundus does not usually exhibit spontaneous electrical oscillations, but the corpus and antrum do exhibit such oscillations. In the intact stomach, the ECA is entrained to a frequency of about 5 cpm (about 3 cpm in humans) throughout the electrically active region with phase lags in both the longitudinal and circumferential directions [Sarna et al., 1972]. The phase lags decrease distally from corpus to antrum.

There is a marked intrinsic frequency gradient along the axis of the stomach and a slight intrinsic frequency gradient along the circumference. The intrinsic frequency of gastric ECA in isolated circular muscle of the orad and mid-corpus is the highest (about 5 cpm) compared to about 3.5 cpm in the rest of the corpus, and about 0.5 cpm in the antrum. Also, there is an orad to aborad intrinsic gradient in resting membrane potential, with the terminal antrum having the most negative resting membrane potential, about 30 mV more negative than the fundal regions [Szurszewski, 1987]. The relatively depolarized state of the fundal muscle may explain its electrical inactivity since the voltage-sensitive ionic channels may be kept in a state of inactivation. Hyperpolarization of the fundus to a transmembrane voltage of –60 mV produces fundal control waves similar to those recorded from mid- and orad corpus.

The ECA in canine stomach was modeled [Sarna et al., 1972] using an array of 13 bidirectionally coupled relaxation oscillators. The model featured (1) an intrinsic frequency decline from corpus to the pylorus and from greater curvature to the lesser curvature, (2) entrainment of all coupled oscillators at a frequency close to the highest intrinsic frequency, and (3) distally decreasing phase lags between the entrained oscillators. A simulated circumferential transection caused the formation of another frequency plateau aboral to the transection. The frequency of the orad plateau remained unaffected while that of the aborad plateau was decreased. This is consistent with the observed experimental data.

The Electrogastrogram

In a similar manner to other electrophysiological measures such as the electrocardiogram (EKG) and the electroencephalogram (EEG), the electrogastrogram (EGG) was identified [Stern and Koch, 1985; Chen and McCallum, 1994]. The EGG is the signal obtained from cutaneous recording of the gastric myoelectrical activity by using surface electrodes placed on the abdomen over the stomach. Although the first EGG was recorded in the early 1920s [Alvarez, 1922], progress *vis-a-vis* clinical applications has been relatively slow, in particular when compared to the progress made in EKG, which also started in the early 1920s. Despite many attempts made over the decades, visual inspection of the EGG signal has not led to the identification of waveform characteristics that would help the clinician to diagnose functional or organic diseases of the stomach. Even the development of techniques such as time-frequency analysis [Qiao et al., 1998] and artificial neural network-based feature extraction [Liang et al., 1997; Wang et al., 1999] for computer analysis of the EGG did not provide *clinically relevant* information about gastric motility disorders. It has been demonstrated that increased EGG frequencies (1) were seen in perfectly healthy subjects [Pffafenbach et al., 1995], and (2) did not always correspond to serosally recorded tachygastria in dogs [Mintchev and Bowes, 1997]. As yet, there is no effective method of detecting a change in the direction or velocity of propagation of gastric ECA from the EGG.

6.5   The Small Intestine

Anatomical Features

The small intestine is a long hollow organ that consists of the duodenum, jejunum, and ileum, respectively. Its length is about 650 cm in humans and 300 cm in dogs. The duodenum extends from the pylorus to the ligament of Treitz (about 30 cm in humans and dogs). In humans, the duodenum forms a C-shaped pattern, with the ligament of Treitz near the corpus of the stomach. In dogs, the duodenum lies along the right side of the peritoneal cavity, with the ligament of Treitz in the pelvis. The duodenum receives pancreatic exocrine secretions and bile. In both humans and dogs, the jejunum consists of the next one third whereas the ileum consists of the remaining two thirds of the intestine. The major differences between the jejunum and ileum are functional in nature, relating to their absorption characteristics and motor control. The majority of sugars, amino acids, lipids, electrolytes, and water are absorbed in the jejunum and proximal ileum, whereas bile acids and vitamin B12 are absorbed in the terminal ileum.

Small Intestinal ECA

In the canine small intestine, the ECA is not entrained throughout the entire length [Diamant and Bortoff, 1969a; Sarna et al., 1971]. However, the ECA exhibits a plateau of constant frequency in the proximal region whereby there is a distal increase in phase lag. The frequency plateau (of about 20 cpm) extends over the entire duodenum and part of the jejunum. There is a marked intrinsic frequency gradient in the longitudinal direction with the highest intrinsic frequency being less than the plateau frequency. When the small intestine was transected *in vivo* into small segments (15 cm long), the intrinsic frequency of the ECA in adjacent segments tended to decrease aborally in an exponential manner [Sarna et al., 1971]. A single transection of the duodenum caused the formation of another frequency plateau aboral to the transection. The ECA frequency in the orad plateau was generally unaffected, while that in the aborad plateau was decreased [Diamant and Bortoff, 1969b; Sarna et al., 1971]. The frequency of the aborad plateau was either higher than or equal to the highest intrinsic frequency distal to the transection, depending on whether the transection of the duodenum was either above or below the region of the bile duct [Diamant and Bortoff, 1969b].

The ECA in canine small intestine was modeled using a chain of 16 bidirectionally coupled relaxation oscillators [Sarna et al., 1971]. Coupling was not uniform along the chain, since the proximal oscillators were strongly coupled and the distal oscillators were weakly coupled. The model featured (1) an exponential intrinsic frequency decline along the chain, (2) a frequency plateau which is higher than the highest intrinsic frequency, and (3) a temporal variation of the frequencies distal to the frequency plateau region. A simulated transection in the frequency plateau region caused the formation of another frequency plateau aboral to the transection, such that the frequency of the orad plateau was unchanged whereas the frequency of the aborad plateau decreased.

The ECA in human small intestine was modeled using a chain of 100 bidirectionally coupled relaxation oscillators [Robertson-Dunn and Linkens, 1976]. Coupling was nonuniform and asymmetrical. The model featured (1) a piecewise linear decline in intrinsic frequency along the chain, (2) a piecewise linear decline in coupling similar to that of the intrinsic frequency, (3) forward coupling which is stronger than backward coupling, and (4) a frequency plateau in the proximal region which is higher than the highest intrinsic frequency in the region.

Small Intestinal MMC

The MMCs in canine small intestine have been observed in intrinsically isolated segments [Sarna et al., 1981; 1983], even after the isolated segment has been stripped of all extrinsic innervation [Sarr and Kelly, 1981] or removed in continuity with the remaining gut as a Thiry Vella loop [Itoh et al., 1981]. This intrinsic mechanism is able to function independently of extrinsic innervation since vagotomy [Weisbrodt et al., 1975; Ruckebusch and Bueno, 1977] does not hinder the initiation of the MMC. The initiation of the small intestinal MMC is controlled by integrative networks within the intrinsic plexuses utilizing nicotinic and muscarinic cholinergic receptors [Ormsbee et al., 1979; El-Sharkawy et al., 1982].

When the canine small intestine was transected into four equal strips [Sarna et al., 1981, 1983], it was found that each strip was capable of generating an independent MMC that would appear to propagate from the proximal to the distal part of each segment. This suggested that the MMC can be modeled by a chain of coupled relaxation oscillators. The average intrinsic periods of the MMC for the four segments were reported to be 106.2, 66.8, 83.1, and 94.8 min, respectively. The segment containing the duodenum had the longest period, while the subsequent segment containing the jejunum had the shortest period. However, in the intact small intestine, the MMC starts in the duodenum and not the jejunum. Bardakjian et al. [1981, 1984] have demonstrated that both the intrinsic frequency gradients and resting level gradients have major roles in the entrainment of a chain of coupled oscillators. In modeling the small intestinal MMC with a chain of four coupled oscillators, it was necessary to include a gradient in the intrinsic resting levels of the MMC oscillators (with the proximal oscillator having the lowest resting level) in order to entrain the oscillators and allow the proximal oscillator to behave as the leading oscillator [Bardakjian and Ahmed, 1992].

6.6　The Colon

Anatomical Features

In humans, the colon is about 100 cm in length. The ileum joins the colon approximately 5 cm from its end, forming the cecum which has a worm-like appendage, the appendix. The colon is sacculated, and the longitudinal smooth muscle is concentrated in three bands (the taeniae). It lies in front of the small intestine against the abdominal wall and it consists of the ascending (on the right side), transverse (across the lower stomach), and descending (on the left side) colon. The descending colon becomes the sigmoid colon in the pelvis as it runs down and forward to the rectum. Major functions of the colon are (1) to absorb water, certain electrolytes, short-chain fatty acids, and bacterial metabolites; (2) to slowly propel its luminal contents in the caudad direction; (3) to store the residual matter in the distal region; and (4) to rapidly move its contents in the caudad direction during mass movements [Sarna, 1991]. In dogs, the colon is about 45 cm in length and the cecum has no appendage. The colon is not sacculated, and the longitudinal smooth muscle coat is continuous around the circumference [Miller et al., 1968]. It lies posterior to the small intestine and it consists mainly of ascending and descending segments with a small transverse segment. However, functionally it is assumed to consist of three regions, each of about 15 cm in length, representing the ascending, transverse, and descending colon, respectively.

Colonic ECA

In the human colon, the ECA is almost completely phase-unlocked between adjacent sites as close as 1 to 2 cm apart, and its frequency (about 3 to 15 cpm) and amplitude at each site vary with time [Sarna et al., 1980]. This results in short duration contractions that are also disorganized in time and space. The disorganization of ECA and its associated contractions is consistent with the colonic function of extensive mixing, kneading, and slow net distal propulsion [Sarna, 1991]. In the canine colon, the reports about the intrinsic frequency gradient were conflicting [Vanasin et al., 1974; Shearin et al., 1978; El-Sharkawy, 1983].

The human colonic ECA was modeled [Bardakjian and Sarna, 1980] using a tubular structure of 99 bidirectionally coupled nonlinear oscillators arranged in 33 parallel rings where each ring contained 3 oscillators. Coupling was nonuniform and it increased in the longitudinal direction. The model featured (1) no phase-locking in the longitudinal or circumferential directions, (2) temporal and spatial variation of the frequency profile with large variations in the proximal and distal regions and small variations in the middle region, and (3) waxing and waning of the amplitudes of the ECA, which was more pronounced in the proximal and distal regions. The model demonstrated that the "silent periods" occurred because of the interaction between oscillators and they did not occur when the oscillators were uncoupled. The model was further refined [Bardakjian et al., 1990] such that when the ECA amplitude exceeded an excitation threshold, a burst of ERA was exhibited. The ERA bursts occurred in a seemingly random manner in adjacent sites because (1) the ECA was not phase-locked, and (2) the ECA amplitudes and waveshapes varied in a seemingly random manner.

6.7 Epilogue

The ECA in stomach, small intestine, and colon behaves like the outputs of a population of coupled nonlinear oscillators. The populations in the stomach and the proximal small intestine are entrained, whereas those in the distal small intestine and colon are not entrained. There are distinct intrinsic frequency gradients in the stomach and small intestine but their profile in the colon is ambiguous.

The applicability of modeling of gastrointestinal ECA by coupled nonlinear oscillators has been reconfirmed [Daniel et al., 1994], and a novel nonlinear oscillator, the mapped clock oscillator, was proposed [Bardakjian and Diamant, 1994] for modeling the cellular ECA. The oscillator consists of two coupled components: a clock which represents the interstitial cells of Cajal, and a transformer that represents the smooth muscle transmembrane ionic transport mechanisms [Skinner and Bardakjian, 1991]. Such a model accounts for the mounting evidence supporting the role of the interstitial cells of Cajal as a pacemaker for the smooth muscle transmembrane voltage oscillations [Hara et al., 1986; Suzuki et al., 1986; Barajas-Lopez et al., 1989; Serio et al., 1991; Sanders, 1996].

Modeling of the gastrointestinal ECA by populations of coupled nonlinear oscillators [Bardakjian, 1987] suggests that gastrointestinal motility disorders associated with abnormal ECA can be effectively treated by (1) electronic pacemakers to coordinate the oscillators, (2) surgical interventions to remove regional ectopic foci, and (3) pharmacotherapy to stimulate the oscillators. Electronic pacing has been demonstrated in canine stomach [Kelly and LaForce, 1972; Sarna and Daniel, 1973; Bellahsene et al.,

1992] and small intestine [Sarna and Daniel, 1975c; Becker et al., 1983]. Also, pharmacotherapy with prokinetic drugs such as Domperidone and Cisapride has demonstrated improvements in the coordination of the gastric oscillators.

Acknowledgments

The author would like to thank his colleagues, Dr. Sharon Chung and Dr. Karen Hall, for providing biological insight.

References

Alvarez, W.C. and Mahoney, L.J. 1922. Action current in stomach and intestine. *Am. J. Physiol.*, 58:476-493.

Alvarez, W.C. 1922. The electrogastrogram and what it shows. *J. Am. Med. Assoc.*, 78:1116-1119.

Alvarez, W.C. and Mahoney, L.J. 1923. The relations between gastric and duodenal peristalsis. *Am. J. Physiol.*, 64:371-386.

Barajas-Lopez, C., Berezin, I., Daniel, E.E., and Huizinga, J.D. 1989. Pacemaker activity recorded in interstitial cells of Cajal of the gastrointestinal tract. *Am. J. Physiol.*, 257:C830-C835.

Bardakjian, B.L. and Sarna, S.K. 1980. A computer model of human colonic electrical control activity (ECA). *IEEE Trans. Biomed. Eng.*, 27:193-202.

Bardakjian, B.L. and Sarna, S.K. 1981. Mathematical investigation of populations of coupled synthesized relaxation oscillators representing biological rhythms. *IEEE Trans. Biomed. Eng.*, 28:10-15.

Bardakjian, B.L., El-Sharkawy, T.Y., and Diamant, N.E. 1984. Interaction of coupled nonlinear oscillators having different intrinsic resting levels. *J. Theor. Biol.*, 106:9-23.

Bardakjian, B.L. 1987. Computer models of gastrointestinal myoelectric activity. *Automedica*, 7:261-276.

Bardakjian, B.L., Sarna, S.K., and Diamant, N.E. 1990. Composite synthesized relaxation oscillators: Application to modeling of colonic ECA and ERA. *Gastrointest. J. Motil.*, 2:109-116.

Bardakjian, B.L. and Ahmed, K. 1992. Is a peripheral pattern generator sufficient to produce both fasting and postprandial patterns of the migrating myoelectric complex (MMC)? *Dig. Dis. Sci.*, 37:986.

Bardakjian, B.L. and Diamant, N.E. 1994. A mapped clock oscillator model for transmembrane electrical rhythmic activity in excitable cells. *J. Theor. Biol.*, 166:225-235.

Bass, P. 1965. Electric activity of smooth muscle of the gastrointestinal tract. *Gastroenterology*, 49:391-394.

Beck, I.T., McKenna, R.D., Peterfy, G., Sidorov, J., and Strawczynski, H. 1965. Pressure studies in the normal human jejunum. *Am. J. Dig. Dis.*, 10:437-448.

Becker, J.M., Sava, P., Kelly, K.A., and Shturman, L. 1983. Intestinal pacing for canine postgastrectomy dumping. *Gastroenterology*, 84:383-387.

Bellahsene, B.E., Lind, C.D., Schirmer, B.D., et al. 1992. Acceleration of gastric emptying with electrical stimulation in a canine model of gastroparesis. *Am. J. Physiol.*, 262:G826-G834.

Boldireff, W.N. 1905. Le travail periodique de l'appareil digestif en dehors de la digestion. *Arch. Des. Sci. Biol.*, 11:1-157.

Bozler, E. 1938. Action potentials of visceral smooth muscle. *Am. J. Physiol.*, 124:502-510.

Bozler, E. 1939. Electrophysiological studies on the motility of the gastrointestinal tract. *Am. J. Physiol.*, 127:301-307.

Bozler, E. 1941. Action potentials and conduction of excitation in muscle. *Biol. Symposia*, 3:95-110.

Bulbring, E., Burnstock G., and Holman, M.E. 1958. Excitation and conduction in the smooth muscle of the isolated taenia coli of the guinea pig. *J. Physiol.*, 142:420-437.

Burnstock, G., Holman, M.E., and Prosser, C.L. 1963. Electrophysiology of smooth muscle. *Physiol. Rev.*, 43:482-527.

Cannon, W.B. and Washburn, A.L. 1912. An explanation of hunger. *Am. J. Physiol.*, 29:441-454.

Carlson, G.M., Bedi, B.S., and Code, C.F. 1972. Mechanism of propagation of intestinal interdigestive myoelectric complex. *Am. J. Physiol.*, 222:1027-1030.

Chen, J.Z. and McCallum, R.W. 1994. *Electrogastrography: Principles and Applications.* Raven Press, New York.

Christensen, J. 1975. Myoelectric control of the colon. *Gastroenterology,* 68:601-609.

Code, C.F. and Marlett, J.A. 1975. The interdigestive myoelectric complex of the stomach and small bowel of dogs. *J. Physiol.,* 246:289-309.

Daniel, E.E. and Chapman, K.M. 1963. Electrical activity of the gastrointestinal tract as an indication of mechanical activity. *Am. J. Dig. Dis.,* 8:54-102.

Daniel, E.E. 1975. Electrophysiology of the colon. *Gut,* 16:298-329.

Daniel, E.E. and Sakai, Y. 1984. Structural basis for function of circular muscle of canine corpus. *Can. J. Physiol. Pharmacol.,* 62:1304-1314.

Daniel, E.E., Bardakjian, B.L., Huizinga, J.D., and Diamant, N.E. 1994. Relaxation oscillators and core conductor models are needed for understanding of GI electrical activities. *Am. J. Physiol.,* 266:G339-G349.

Diamant, N.E. and Bortoff, A. 1969a. Nature of the intestinal slow wave frequency gradient. *Am. J. Physiol.,* 216:301-307.

Diamant, N.E. and Bortoff, A. 1969b. Effects of transection on the intestinal slow wave frequency gradient. *Am. J. Physiol.,* 216:734-743.

Douglas, D.M. 1949. The decrease in frequency of contraction of the jejunum after transplantation to the ileum. *J. Physiol.,* 110:66-75.

Duthie, H.L. 1974. Electrical activity of gastrointestinal smooth muscle. *Gut,* 15:669-681.

El-Sharkawy, T.Y., Markus, H., and Diamant, N.E. 1982. Neural control of the intestinal migrating myoelectric complex: A pharmacological analysis. *Can. J. Physiol. Pharm.,* 60:794-804.

El-Sharkawy, T.Y. 1983. Electrical activity of the muscle layers of the canine colon. *J. Physiol.,* 342:67-83.

Fleckenstein, P. 1978. Migrating electrical spike activity in the fasting human small intestine. *Dig. Dis. Sci.,* 23:769-775.

Gillespie, J.S. 1962. The electrical and mechanical responses of intestinal smooth muscle cells to stimulation of their extrinsic parasympathetic nerves. *J. Physiol.,* 162:76-92.

Grivel, M.L. and Ruckebusch, Y. 1972. The propagation of segmental contractions along the small intestine. *J. Physiol.,* 277:611-625.

Hara, Y.M., Kubota, M., and Szurszewski, J.H. 1986. Electrophysiology of smooth muscle of the small intestine of some mammals. *J. Physiol.,* 372:501-520.

Itoh, Z., Honda, R., Aizawa, I., Takeuchi, S., Hiwatashi, K., and Couch, E.F. 1978. Interdigestive motor activity of the lower esophageal sphincter in the conscious dog. *Dig. Dis. Sci.,* 23:239-247.

Itoh, Z., Aizawa, I., and Takeuchi, S. 1981. Neural regulation of interdigestive motor activity in canine jejunum. *Am. J. Physiol.,* 240:G324-G330.

Jacoby, H.I., Bass, P., and Bennett, D.R. 1963. *In vivo* extraluminal contractile force transducer for gastrointestinal muscle. *J. Appl. Physiol.,* 18:658-665.

Kelly, K.A. and LaForce, R.C. 1972. Pacing the canine stomach with electric stimulation. *Am. J. Physiol.,* 222:588-594.

Kerlin, P. and Phillips, S. 1982. The variability of motility of the ileum and jejunum in healthy humans. *Gastroenterology,* 82:694-700.

Liang, J., Cheung, J.Y., and Chen, J.D.Z. 1997. Detection and deletion of motion artifacts in electrogastrogram using feature analysis and neural networks. *Ann. Biomed. Eng.,* 25:850-857.

Linkens, D.A., Taylor, I., and Duthie, H.L. 1976. Mathematical modeling of the colorectal myoelectrical activity in humans. *IEEE Trans. Biomed. Eng.,* 23:101-110.

Milton, G.W. and Smith, A.W.M. 1956. The pacemaking area of the duodenum. *J. Physiol.,* 132:100-114.

Miller, M.E., Christensen, G.C., and Evans, H.E. 1968. *Anatomy of the Dog,* Saunders, Philadelphia.

Mintchev, M.P. and Bowes, K.L. 1997. Do increased electrogastrographic frequencies always correspond to internal tachygastria? *Ann. Biomed. Eng.,* 25:1052-1058.

Morat, J.P. 1882. Sur l'innervation motrice de l'estomac. *Lyon. Med.,* 40:289-296.

Morat, J.P. 1893. Sur quelques particularites de l'innervation motrice de l'estomac et de l'intestin. *Arch. de Physiol. Norm. et Path.,* 5:142-153.

Nelson, T.S. and Becker, J.C. 1968. Simulation of the electrical and mechanical gradient of the small intestine. *Am. J. Physiol.,* 214:749-757.

Ormsbee, H.S., Telford, G.L., and Mason, G.R. 1979. Required neural involvement in control of canine migrating motor complex. *Am. J. Physiol.,* 237:E451-E456.

Pffafenbach, B., Adamek, R.J., Kuhn, K., and Wegener, M. 1995. Electrogastrography in healthy subjects. Evaluation of normal values: influence of age and gender. *Dig. Dis. Sci.,* 40:1445-1450.

Pick, T.P. and Howden, R. 1977. *Gray's Anatomy,* Bounty Books, New York.

Puestow, C.B. 1933. Studies on the origins of the automaticity of the intestine: The action of certain drugs on isolated intestinal transplants. *Am. J. Physiol.,* 106:682-688.

Qiao, W., Sun, H.H., Chey, W.Y., and Lee, K.Y. 1998. Continuous wavelet analysis as an aid in the representation and interpretation of electrogastrographic signals. *Ann. Biomed. Eng.,* 26:1072-1081.

Rees, W.D.W., Malagelada, J.R., Miller, L.J., and Go, V.L.W. 1982. Human interdigestive and postprandial gastrointestinal motor and gastrointestinal hormone patterns. *Dig. Dis. Sci.,* 27:321-329.

Reinke, D.A., Rosenbaum, A.H., and Bennett, D.R. 1967. Patterns of dog gastrointestinal contractile activity monitored *in vivo* with extraluminal force transducers. *Am. J. Dig. Dis.,* 12:113-141.

Robertson-Dunn, B. and Linkens, D.A. 1974. A mathematical model of the slow wave electrical activity of the human small intestine. *Med. Biol. Eng.,* 12:750-758.

Ruckebusch, Y. and Fioramonti, S. 1975. Electrical spiking activity and propulsion in small intestine in fed and fasted states. *Gastroenterology,* 68:1500-1508.

Ruckebusch, Y. and Bueno, L. 1976. The effects of feeding on the motility of the stomach and small intestine in the pig. *Br. J. Nutr.,* 35:397-405.

Ruckebusch, Y. and Bueno, L. 1977. Migrating myoelectrical complex of the small intestine. *Gastroenterology,* 73:1309-1314.

Sanders, K.M. 1996. A case for interstitial cells of Cajal as pacemakers and mediators of neurotransmission in the gastrointestinal tract. *Gastroenterology,* 111(2):492-515.

Shearin, N.L., Bowes, K.L. and Kingma, Y.J. 1978. *In vitro* electrical activity in canine colon. *Gut,* 20:780-786.

Sarna, S.K., Daniel, E.E., and Kingma, Y.J. 1971. Simulation of slow wave electrical activity of small intestine. *Am. J. Physiol.,* 221:166-175.

Sarna, S.K., Daniel, E.E., and Kingma, Y.J. 1972. Simulation of the electrical control activity of the stomach by an array of relaxation oscillators. *Am. J. Dig. Dis.,* 17:299-310.

Sarna, S.K. and Daniel, E.E. 1973. Electrical stimulation of gastric electrical control activity. *Am. J. Physiol.,* 225:125-131.

Sarna, S.K. 1975a. Models of smooth muscle electrical activity. In *Methods in Pharmacology,* E.E. Daniel and D.M. Paton, Eds., Plenum Press, New York, 519-540.

Sarna, S.K. 1975b. Gastrointestinal electrical activity: terminology. *Gastroenterology,* 68:1631-1635.

Sarna, S.K. and Daniel, E.E. 1975c. electrical stimulation of small intestinal electrical control activity. *Gastroenterology,* 69:660-667.

Sarna, S.K., Bardakjian, B.L., Waterfall, W.E., and Lind, J.F. 1980. Human colonic electrical control activity (ECA). *Gastroenterology,* 78:1526-1536.

Sarna, S.K., Stoddard, C., Belbeck L., and McWade D. 1981. Intrinsic nervous control of migrating myoelectric complexes. *Am. J. Physiol.,* 241:G16-G23.

Sarna, S., Condon, R.E., and Cowles, V. 1983. Enteric mechanisms of initiation of migrating myoelectric complexes in dogs. *Gastroenterology,* 84:814-822.

Sarna, S.K. 1991. Physiology and pathophysiology of colonic motor activity. *Dig. Dis. Sci.,* 6:827-862.

Sarr M.G. and Kelly, K.A. 1981. Myoelectric activity of the autotransplanted canine jejunoileum. *Gastroenterology,* 81:303-310.

Serio, R., Barajas-Lopez, C., Daniel, E.E., Berezin, I., and Huizinga, J.D. 1991. Pacemaker activity in the colon: Role of interstitial cells of Cajal and smooth muscle cells. *Am. J. Physiol.,* 260:G636-G645.

Stanciu, C. and Bennett, J.R. 1975. The general pattern of gastroduodenal motility: 24 hour recordings in normal subjects. *Rev. Med. Chir. Soc. Med. Nat. Iasi.,* 79:31-36.

Skinner, F.K. and Bardakjian, B.L. 1991. A barrier kinetic mapping unit. Application to ionic transport in gastric smooth muscle. *Gastrointest. J. Motil.,* 3:213-224.

Stern, R.M. and Koch, K.L. 1985. *Electrogastrography: Methodology, Validation, and Applications.* Praeger Publishers, New York.

Summers, R.W., Anuras, S., and Green, J. 1982. Jejunal motility patterns in normal subjects and symptomatic patients with partial mechanical obstruction or pseudo-obstruction. In *Motility of the Digestive Tract,* M. Weinbeck, Ed., Raven Press, New York, 467-470.

Suzuki, N., Prosser, C.L., and Dahms, V., 1986. Boundary cells between longitudinal and circular layers: Essential for electrical slow waves in cat intestine. *Am. J. Physiol.,* 280:G287-G294.

Szurszewski, J.H. 1969. A migrating electric complex of the canine small intestine. *Am. J. Physiol.,* 217:1757-1763.

Szurszewski, J.H. 1987. Electrical basis for gastrointestinal motility. In *Physiology of the Gastrointestinal Tract,* L.R. Johnson, Ed., Raven Press, New York, chap. 12.

Thompson, D.G., Wingate, D.L., Archer, L., Benson, M.J., Green, W.J., and Hardy, R.J. 1980. Normal patterns of human upper small bowel motor activity recorded by prolonged radiotelemetry. *Gut,* 21:500-506.

Vanasin, B., Ustach, T.J., and Schuster, M.M. 1974. Electrical and motor activity of human and dog colon in vitro. *Johns Hopkins Med. J.,* 134:201-210.

Vantrappen, G., Janssens, J.J., Hellemans, J. and Ghoos, Y. 1977. The interdigestive motor complex of normal subjects and patients with bacterial overgrowth of the small intestine. *J. Clin. Invest.,* 59:1158-1166.

Wang, Z., He, Z., and Chen, J.D.Z. 1999. Filter banks and neural network-based feature extraction and automatic classification of electrogastrogram. *Ann. Biomed. Eng.,* 27:88-95.

Weisbrodt, N.W., Copeland, E.M., Moore, E.P., Kearly, K.W., and Johnson, L.R. 1975. Effect of vagotomy on electrical activity of the small intestine of the dog. *Am. J. Physiol.,* 228:650-654.

Wheelon, H., and Thomas, J.E. 1921. Rhythmicity of the pyloric sphincter. *Am. J. Physiol.,* 54:460-473.

Wheelon, H., and Thomas, J.E. 1922. Observations on the motility of the duodenum and the relation of duodenal activity to that of the pars pylorica. *Am. J. Physiol.,* 59:72-96.

7
Respiratory System

Arthur T. Johnson
University of Maryland

Christopher G. Lausted
University of Maryland

Joseph D. Bronzino
Trinity College/The Biomedical Engineering Alliance and Consortium (BEACON)

As functioning units, the lung and heart are usually considered a single complex organ, but because these organs contain essentially two compartments — one for blood and one for air — they are usually separated in terms of the tests conducted to evaluate heart or pulmonary function. This chapter focuses on some of the physiologic concepts responsible for normal function and specific measures of the lung's ability to supply tissue cells with enough oxygen while removing excess carbon dioxide.

7.1   Respiration Anatomy

The respiratory system consists of the lungs, conducting airways, pulmonary vasculature, respiratory muscles, and surrounding tissues and structures (Fig. 7.1). Each plays an important role in influencing respiratory responses.

Lungs

There are two lungs in the human chest; the right lung is composed of three incomplete divisions called *lobes,* and the left lung has two, leaving room for the heart. The right lung accounts for 55% of total gas volume and the left lung for 45%. Lung tissue is spongy because of the very small (200 to 300×10^{-6} m diameter in normal lungs at rest) gas-filled cavities called *alveoli,* which are the ultimate structures for gas exchange. There are 250 million to 350 million alveoli in the adult lung, with a total alveolar surface area of 50 to 100 m^2 depending on the degree of lung inflation [Johnson, 1991].

Conducting Airways

Air is transported from the atmosphere to the alveoli beginning with the oral and nasal cavities, through the pharynx (in the throat), past the glottal opening, and into the trachea or windpipe. Conduction of air begins at the larynx, or voice box, at the entrance to the trachea, which is a fibromuscular tube 10 to 12 cm in length and 1.4 to 2.0 cm in diameter [Kline, 1976]. At a location called the *carina,* the trachea terminates and divides into the left and right bronchi. Each bronchus has a discontinuous cartilaginous support in its wall. Muscle fibers capable of controlling airway diameter are incorporated into the walls

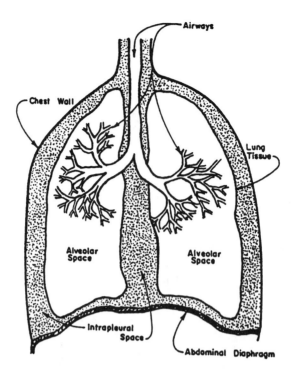

FIGURE 7.1   Schematic representation of the respiratory system.

of the bronchi, as well as in those of air passages closer to the alveoli. Smooth muscle is present throughout the respiratory bronchiolus and alveolar ducts but is absent in the last alveolar duct, which terminates in one to several alveoli. The alveolar walls are shared by other alveoli and are composed of highly pliable and collapsible squamous epithelium cells.

The bronchi subdivide into subbronchi, which further subdivide into bronchioli, which further subdivide, and so on, until finally reaching the alveolar level. Table 7.1 provides a description and dimensions of the airways of adult humans. A model of the geometric arrangement of these air passages is presented in Fig. 7.2. It will be noted that each airway is considered to branch into two subairways. In the adult human there are considered to be 23 such branchings, or generations, beginning at the trachea and ending in the alveoli.

Movement of gases in the respiratory airways occurs mainly by bulk flow (convection) throughout the region from the mouth to the nose to the fifteenth generation. Beyond the fifteenth generation, gas diffusion is relatively more important. With the low gas velocities that occur in diffusion, dimensions of the space over which diffusion occurs (alveolar space) must be small for adequate oxygen delivery into the walls; smaller alveoli are more efficient in the transfer of gas than are larger ones. Thus animals with high levels of oxygen consumption are found to have smaller-diameter alveoli compared with animals with low levels of oxygen consumption.

Alveoli

Alveoli are the structures through which gases diffuse to and from the body. To ensure gas exchange occurs efficiently, alveolar walls are extremely thin. For example, the total tissue thickness between the inside of the alveolus to pulmonary capillary blood plasma is only about 0.4×10^{-6} m. Consequently, the principal barrier to diffusion occurs at the plasma and red blood cell level, not at the alveolar membrane [Ruch and Patton, 1966].

Molecular diffusion within the alveolar volume is responsible for mixing of the enclosed gas. Due to small alveolar dimensions, complete mixing probably occurs in less than 10 ms, fast enough that alveolar mixing time does not limit gaseous diffusion to or from the blood [Astrand and Rodahl, 1970].

TABLE 7.1 Classification and Approximate Dimensions of Airways of Adult Human Lung (Inflated to about 3/4 of TLC)*

Common Name	Numerical Order of Generation	Number of Each	Diameter, mm	Length, mm	Total Cross-Sectional Area, cm²	Description and Comment
Trachea	0	1	18	120	2.5	Main cartilaginous airway; partly in thorax.
Main bronchus	1	2	12	47.6	2.3	First branching of airway; one to each lung; in lung root; cartilage.
Lobar bronchus	2	4	8	19.0	2.1	Named for each lobe; cartilage.
Segmental bronchus	3	8	6	7.6	2.0	Named for radiographical and surgical anatomy; cartilage.
Subsegmental bronchus	4	16	4	12.7	2.4	Last generally named bronchi; may be referred to as medium-sized bronchi; cartilage.
Small bronchi	5–10	1,024[†]	1.3[†]	4.6[†]	13.4[†]	Not generally named; contain decreasing amounts of cartilage. Beyond this level airways enter the lobules as defined by a strong elastic lobular limiting membrane.
Bronchioles	11–13	8,192[†]	0.8[†]	2.7[†]	44.5[†]	Not named; contain no cartilage, mucus-secreting elements, or cilia. Tightly embedded in lung tissue.
Terminal bronchioles	14–15	32,768[†]	0.7[†]	2.0[†]	113.0[†]	Generally 2 or 3 orders so designated; morphology not significantly different from orders 11–13.
Respiratory bronchioles	16–18	262,144[†]	0.5[†]	1.2[†]	534.0[†]	Definite class; bronchiolar cuboidal epithelium present, but scattered alveoli are present giving these airways a gas exchange function. Order 165 often called first-order respiratory bronchiole; 17, second-order; 18, third-order.
Alveolar ducts	19–22	4,194,304[†]	0.4[†]	0.8[†]	5,880.0[†]	No bronchial epithelium; have no surface except connective tissue framework; open into alveoli.
Alveolar sacs	23	8,388,608	0.4	0.6	11,800.0	No reason to assign a special name; are really short alveolar ducts.
Alveoli	24	300,000,000	0.2			Pulmonary capillaries are in the septae that form the alveoli.

* The number of airways in each generation is based on regular dichotomous branching.

† Numbers refer to last generation in each group.

Source: Used with permission from Staub [1963] and Weibel [1963]; adapted by Comroe [1965].

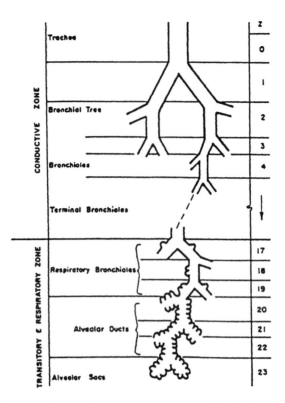

FIGURE 7.2 General architecture of conductive and transitory airways. (Used with permission from Weibel, 1963.) In the conductive zone, air is conducted to and from the lungs while in the respiration zone, gas exchange occurs.

Of particular importance to proper alveolar operation is a thin surface coating of surfactant. Without this material, large alveoli would tend to enlarge and small alveoli would collapse. It is the present view that surfactant acts like a detergent, changing the stress-strain relationship of the alveolar wall and thereby stabilizing the lung [Johnson, 1991].

Pulmonary Circulation

There is no true pulmonary analog to the systemic arterioles, since the pulmonary circulation occurs under relatively low pressure [West, 1977]. Pulmonary blood vessels, especially capillaries and venules, are very thin walled and flexible. Unlike systemic capillaries, pulmonary capillaries increase in diameter, and pulmonary capillaries within alveolar walls separate adjacent alveoli with increases in blood pressure or decreases in alveolar pressure. Flow, therefore, is significantly influenced by elastic deformation. Although pulmonary circulation is largely unaffected by neural and chemical control, it does respond promptly to hypoxia.

There is also a high-pressure systemic blood delivery system to the bronchi that is completely independent of the pulmonary low-pressure ($\sim$3330 N/m^2) circulation in healthy individuals. In diseased states, however, bronchial arteries are reported to enlarge when pulmonary blood flow is reduced, and some arteriovenous shunts become prominent [West, 1977].

Total pulmonary blood volume is approximately 300 to 500 cm^3 in normal adults, with about 60 to 100 cm^3 in the pulmonary capillaries [Astrand and Rodahl, 1970]. This value, however, is quite variable, depending on such things as posture, position, disease, and chemical composition of the blood [Kline, 1976].

Since pulmonary arterial blood is oxygen-poor and carbon dioxide-rich, it exchanges excess carbon dioxide for oxygen in the pulmonary capillaries, which are in close contact with alveolar walls. At rest, the transit time for blood in the pulmonary capillaries is computed as

$$t = V_c / \dot{V}_c ,$$

where t   = blood transmit time, s
   V_c = capillary blood volume, m³
   $\dot{V}_c$ = total capillary blood flow = cardiac output, m³/s,

and is somewhat less than 1 s, while during exercise it may be only 500 ms or even less.

Respiratory Muscles

The lungs fill because of a rhythmic expansion of the chest wall. The action is indirect in that no muscle acts directly on the lung. The diaphragm, the muscular mass accounting for 75% of the expansion of the chest cavity, is attached around the bottom of the thoracic cage, arches over the liver, and moves downward like a piston when it contracts. The external intercostal muscles are positioned between the ribs and aid inspiration by moving the ribs up and forward. This, then, increases the volume of the thorax. Other muscles are important in the maintenance of thoracic shape during breathing. (For details, see Ruch and Patton [1966] and Johnson [1991]).

Quiet expiration is usually considered to be passive; i.e., pressure to force air from the lungs comes from elastic expansion of the lungs and chest wall. During moderate to severe exercise, the abdominal and internal intercostal muscles are very important in forcing air from the lungs much more quickly than would otherwise occur. Inspiration requires intimate contact between lung tissues, pleural tissues (the pleura is the membrane surrounding the lungs), and chest wall and diaphragm. This is accomplished by reduced intrathoracic pressure (which tends toward negative values) during inspiration.

Viewing the lungs as an entire unit, one can consider the lungs to be elastic sacs within an air-tight barrel — the thorax — which is bounded by the ribs and the diaphragm. Any movement of these two boundaries alters the volume of the lungs. The normal breathing cycle in humans is accomplished by the active contraction of the inspiratory muscles, which enlarges the thorax. This enlargement lowers intrathoracic and interpleural pressure even further, pulls on the lungs, and enlarges the alveoli, alveolar ducts, and bronchioli, expanding the alveolar gas and decreasing its pressure below atmospheric. As a result, air at atmospheric pressure flows easily into the nose, mouth, and trachea.

7.2   Lung Volumes and Gas Exchange

Of primary importance to lung functioning is the movement and mixing of gases within the respiratory system. Depending on the anatomic level under consideration, gas movement is determined mainly by diffusion or convection.

Without the thoracic musculature and rib cage, as mentioned above, the barely inflated lungs would occupy a much smaller space than they occupy *in situ*. However, the thoracic cage holds them open. Conversely, the lungs exert an influence on the thorax, holding it smaller than should be the case without the lungs. Because the lungs and thorax are connected by tissue, the volume occupied by both together is between the extremes represented by relaxed lungs alone and thoracic cavity alone. The resting volume V_R, then, is that volume occupied by the lungs with glottis open and muscles relaxed.

Lung volumes greater than resting volume are achieved during inspiration. Maximum inspiration is represented by *inspiratory reserve volume* (IRV). IRV is the maximum additional volume that can be accommodated by the lung at the end of inspiration. Lung volumes less than resting volume do not normally occur at rest but do occur during exhalation while exercising (when exhalation is active). Maximum additional expiration, as measured from lung volume at the end of expiration, is called *expiratory reserve volume* (ERV). *Residual volume* is the amount of gas remaining in the lungs at the end of maximal expiration.

Tidal volume V_T is normally considered to be the volume of air entering the nose and mouth with each breath. Alveolar ventilation volume, the volume of fresh air that enters the alveoli during each breath,

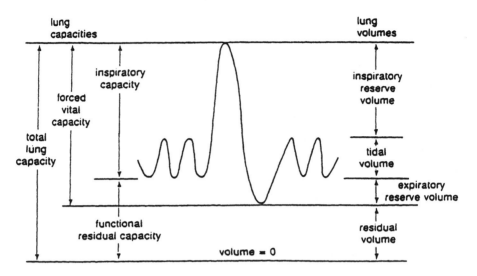

FIGURE 7.3   Lung capacities and lung volumes.

is always less than tidal volume. The extent of this difference in volume depends primarily on the *anatomic dead space*, the 150- to 160-ml internal volume of the conducting airway passages. The term *dead* is quite appropriate, since it represents wasted respiratory effort; i.e., no significant gas exchange occurs across the thick walls of the trachea, bronchi, and bronchiolus. Since normal tidal volume at rest is usually about 500 ml of air per breath, one can easily calculate that because of the presence of this dead space, about 340 to 350 ml of fresh air actually penetrates the alveoli and becomes involved in the gas exchange process. An additional 150 to 160 ml of stale air exhaled during the previous breath is also drawn into the alveoli.

The term *volume* is used for elemental differences of lung volume, whereas the term *capacity* is used for combination of lung volumes. Figure 7.3 illustrates the interrelationship between each of the following lung volumes and capacities:

1. *Total lung capacity* (TLC): The amount of gas contained in the lung at the end of maximal inspiration.
2. *Forced vital capacity* (FVC): The maximal volume of gas that can be forcefully expelled after maximal inspiration.
3. *Inspiratory capacity* (IC): The maximal volume of gas that can be inspired from the resting expiratory level.
4. *Functional residual capacity* (FRC): The volume of gas remaining after normal expiration. It will be noted that functional residual capacity (FRC) is the same as the resting volume. There is a small difference, however, between resting volume and FRC because FRC is measured while the patient breathes, whereas resting volume is measured with no breathing. FRC is properly defined only at end-expiration at rest and not during exercise.

These volumes and specific capacities, represented in Fig. 7.3, have led to the development of specific tests (that will be discussed below) to quantify the status of the pulmonary system. Typical values for these volumes and capacities are provided in Table 7.2.

7.3   Perfusion of the Lung

For gas exchange to occur properly in the lung, air must be delivered to the alveoli via the conducting airways, gas must diffuse from the alveoli to the capillaries through extremely thin walls, and the same gas must be removed to the cardiac atrium by blood flow. This three-step process involves (1) alveolar

TABLE 7.2   Typical Lung Volumes for Normal, Healthy Males

Lung Volume	Normal Values	
Total lung capacity (TLC)	6.0×10^{-3} m^3	(6,000 cm^3)
Residual volume (RV)	1.2×10^{-3} m^3	(1,200 cm^3)
Vital capacity (VC)	4.8×10^{-3} m^3	(4,800 cm^3)
Inspiratory reserve volume (IRV)	3.6×10^{-3} m^3	(3,600 cm^3)
Expiratory reserve volume (ERV)	1.2×10^{-3} m^3	(1,200 cm^3)
Functional residual capacity (FRC)	2.4×10^{-3} m^3	(2,400 cm^3)
Anatomic dead volume (V_D)	1.5×10^{-4} m^3	(150 cm^3)
Upper airways volume	8.0×10^{-5} m^3	(80 cm^3)
Lower airways volume	7.0×10^{-5} m^3	(70 cm^3)
Physiologic dead volume (V_D)	1.8×10^{-4} m^3	(180 cm^3)
Minute volume ($\dot{V}_e$) at rest	1.0×10^{-4} m^3/s	(6,000 cm^3/min)
Respiratory period (T) at rest	4s	
Tidal volume (V_T) at rest	4.0×10^{-4} m^3	(400 cm^3)
Alveolar ventilation volume (V_A) at rest	2.5×10^{-4} m^3	(250 cm^3)
Minute volume during heavy exercise	1.7×10^{-3} m^3/s	(10,000 cm^3/min)
Respiratory period during heavy exercise	1.2 s	
Tidal volume during heavy exercise	2.0×10^{-3} m^3	(2,000 cm^3)
Alveolar ventilation volume during exercise	1.8×10^{-3} m^3	(1,820 cm^3)

Source: Adapted and used with permission from Forster et al. [1986].

ventilation, (2) the process of diffusion, and (3) ventilatory perfusion, which involves pulmonary blood flow. Obviously, an alveolus that is ventilated but not perfused cannot exchange gas. Similarly, a perfused alveolus that is not properly ventilated cannot exchange gas. The most efficient gas exchange occurs when ventilation and perfusion are matched.

There is a wide range of ventilation-to-perfusion ratios that naturally occur in various regions of the lung [Johnson, 1991]. Blood flow is somewhat affected by posture because of the effects of gravity. In the upright position, there is a general reduction in the volume of blood in the thorax, allowing for larger lung volume. Gravity also influences the distribution of blood, such that the perfusion of equal lung volumes is about five times greater at the base compared with the top of the lung [Astrand and Rodahl, 1970]. There is no corresponding distribution of ventilation; hence the ventilation-to-perfusion ratio is nearly five times smaller at the top of the lung (Table 7.3). A more uniform ventilation-to-perfusion ratio is found in the supine position and during exercise [Jones, 1984b].

Blood flow through the capillaries is not steady. Rather, blood flows in a halting manner and may even be stopped if intraalveolar pressure exceeds intracapillary blood pressure during diastole. Mean blood flow is not affected by heart rate [West, 1977], but the highly distensible pulmonary blood vessels admit more blood when blood pressure and cardiac output increase. During exercise, higher pulmonary blood pressures allow more blood to flow through the capillaries. Even mild exercise favors more uniform perfusion of the lungs [Astrand and Rodahl, 1970]. Pulmonary artery systolic pressures increases from 2670 N/m^2 (20 mmHg) at rest to 4670 N/m^2 (35 mmHg) during moderate exercise to 6670 N/m^2 (50 mmHg) at maximal work [Astrand and Rodahl, 1970].

7.4   Gas Partial Pressures

The primary purpose of the respiratory system is gas exchange. In the gas-exchange process, gas must diffuse through the alveolar space, across tissue, and through plasma into the red blood cell, where it finally chemically joins to hemoglobin. A similar process occurs for carbon dioxide elimination.

As long as intermolecular interactions are small, most gases of physiologic significance can be considered to obey the ideal gas law:

$$pV = nRT,$$

TABLE 7.3   Ventilation-to-Perfusion Ratios from the Top to Bottom of the Lung of a Normal Man in the Sitting Position

Percent Lung Volume, %	Alveolar Ventilation Rate, cm³/s	Perfusion Rate, cm³/s	Ventilation-to-Perfusion Ratio
	Top		
7	4.0	1.2	3.3
8	5.5	3.2	1.8
10	7.0	5.5	1.3
11	8.7	8.3	1.0
12	9.8	11.0	0.90
13	11.2	13.8	0.80
13	12.0	16.3	0.73
13	13.0	19.2	0.68
	Bottom		
13	13.7	21.5	0.63
100	84.9	100.0	

Source: Used with permission from West [1962].

where p = pressure, N/m²
       V = volume of gas, m³
       n = number of moles, mol
       R = gas constant, $(N \times m)/(mol \times K)$
       T = absolute temperature, K.

The ideal gas law can be applied without error up to atmospheric pressure; it can be applied to a mixture of gases, such as air, or to its constituents, such as oxygen or nitrogen. All individual gases in a mixture are considered to fill the total volume and have the same temperature but reduced pressures. The pressure exerted by each individual gas is called the *partial pressure* of the gas.

Dalton's law states that the total pressure is the sum of the partial pressures of the constituents of a mixture:

$$p = \sum_{i=1}^{N} p_i \,,$$

where p_i = partial pressure of the *i*th constituent, N/m²
       N = total number of constituents.

Dividing the ideal gas law for a constituent by that for the mixture gives

$$\frac{P_i V}{PV} = \frac{n_i R_i T}{nRT} \,,$$

so that

$$\frac{p_i}{p} = \frac{n_i R_i}{nR} \,,$$

which states that the partial pressure of a gas may be found if the total pressure, mole fraction, and ratio of gas constants are known. For most respiratory calculations, p will be considered to be the pressure of

TABLE 7.4 Molecular Masses, Gas Constants, and Volume Fractions for Air and Constituents

Constituent	Molecular Mass, kg/mol	Gas Constant, N·m/(mol·K)	Volume Fraction in Air, m³/m³
Air	29.0	286.7	1.0000
Ammonia	17.0	489.1	0.0000
Argon	39.9	208.4	0.0093
Carbon dioxide	44.0	189.0	0.0003
Carbon monoxide	28.0	296.9	0.0000
Helium	4.0	2078.6	0.0000
Hydrogen	2.0	4157.2	0.0000
Nitrogen	28.0	296.9	0.7808
Oxygen	32.0	259.8	0.2095

Note: Universal gas constant is 8314.43 N·m/kg·mol·K.

1 atmosphere, 101 kN/m². Avogadro's principle states that different gases at the same temperature and pressure contain equal numbers of molecules:

$$\frac{V_1}{V_2} = \frac{nR_1}{nR_2} = \frac{R_1}{R_2} .$$

Thus

$$\frac{p_i}{p} = \frac{V_i}{V},$$

where V_i/V is the volume fraction of a constituent in air and is therefore dimensionless. Table 7.4 provides individual gas constants, as well as volume fractions, of constituent gases of air.

Gas pressures and volumes can be measured for many different temperature and humidity conditions. Three of these are body temperature and pressure, saturated (BTPS); ambient temperature and pressure (ATP); and standard temperature and pressure, dry (STPD). To calculate constituent partial pressures at STPD, total pressure is taken as barometric pressure minus vapor pressure of water in the atmosphere:

$$p_i = \left(V_i/V\right)\left(p - pH_2O\right),$$

where p $\quad$ = total pressure, kN/m²
$\quad$ pH_2O = vapor pressure of water in atmosphere, kN/m²,

and V_i/V as a ratio does not change in the conversion process.

Gas volume at STPD is converted from ambient condition volume as

$$V_i = V_{amb}\left[273/\left(273 + \Theta\right)\right]\left[\left(p - pH_2O\right)/101.3\right],$$

where V_i $\quad$ = volume of gas i corrected to STPD, m³
$\quad$ V_{amb} = volume of gas i at ambient temperature and pressure, m³
$\quad$ Θ $\quad$ = ambient temperature, °C
$\quad$ p $\quad$ = ambient total pressure, kN/m²
$\quad$ pH_2O = vapor pressure of water in the air, kN/m².

TABLE 7.5   Gas Partial Pressures (kN/m^2) throughout
the Respiratory and Circulatory Systems

Gas	Inspired Air*	Alveolar Air	Expired Air	Mixed Venous Blood	Arterial Blood	Muscle Tissue
H_2O	—	6.3	6.3	6.3	6.3	6.3
CO_2	0.04	5.3	4.2	6.1	5.3	6.7
O_2	21.2	14.0	15.5	5.3	13.3	4.0
N_2†	80.1	75.7	75.3	76.4	76.4	76.4
Total	101.3	101.3	101.3	94.1	101.3	93.4

*Inspired air considered dry for convenience.
†Includes all other inert components.
Source: Used with permission from Astrand and Rodahl [1970].

Partial pressures and gas volumes may be expressed in BTPS conditions. In this case, gas partial pressures are usually known from other measurements. Gas volumes are converted from ambient conditions by

$$V_i = V_{amb}\left[310/(273+\Theta)\right]\left[(p-pH_2O)/p-6.28\right].$$

Table 7.5 provides gas partial pressure throughout the respiratory and circulatory systems.

7.5   Pulmonary Mechanics

The respiratory system exhibits properties of resistance, compliance, and inertance analogous to the electrical properties of resistance, capacitance, and inductance. Of these, inertance is generally considered to be of less importance than the other two properties.

Resistance is the ratio of pressure to flow:

$$R = p/V ,$$

where R = resistance, $N \times s/m^5$
P = pressure, N/m^2
V = volume flow rate, m^3/s.

Resistance can be found in the conducting airways, in the lung tissue, and in the tissues of the chest wall. Airways exhalation resistance is usually higher than airways inhalation resistance because the surrounding lung tissue pulls the smaller, more distensible airways open when the lung is being inflated. Thus airways inhalation resistance is somewhat dependent on lung volume, and airways exhalation resistance can be very lung volume–dependent [Johnson, 1991]. Respiratory tissue resistance varies with frequency, lung volume, and volume history. Tissue resistance is relatively small at high frequencies but increases greatly at low frequencies, nearly proportional to 1/f. Tissue resistance often exceeds airway resistance below 2 Hz. Lung tissue resistance also increases with decreasing volume amplitude [Stamenovic et al., 1990].

Compliance is the ratio of lung volume to lung pressure:

$$C = V/p ,$$

where C = compliance, m^5/N,
V = lung volume/m^3
P = pressure, N/m^2.

As the lung is stretched, it acts as an expanded balloon that tends to push air out and return to its normal size. The static pressure-volume relationship is nonlinear, exhibiting decreased static compliance at the extremes of lung volume [Johnson, 1991]. As with tissue resistance, dynamic tissue compliance does not remain constant during breathing. Dynamic compliance tends to increase with increasing volume and decrease with increasing frequency [Stamenovic et al., 1990].

Two separate approaches can be used to model lung tissue mechanics. The traditional approach places a linear viscoelastic system in parallel with a plastoelastic system. A linear viscoelastic system consists of ideal resistive and compliant elements and can exhibit the frequency-dependence of respiratory tissue. A plastoelastic system consists of dry-friction elements and compliant elements and can exhibit the volume dependence of respiratory tissue [Hildebrandt, 1970]. An alternate approach is to utilize a nonlinear viscoelastic system that can characterize both the frequency dependence and the volume dependence of respiratory tissue [Suki and Bates, 1991].

Lung tissue hysteresivity relates resistance and compliance:

$$wR = \eta / C_{dyn} ,$$

where ω = frequency, radians/s
R = resistance, $N \times s/m^5$
η = hysteresivity, unitless
C_{dyn} = dynamic compliance, m^5/n.

Hysteresivity, analogous to the structural damping coefficient used in solid mechanics, is an empirical parameter arising from the assumption that resistance and compliance are related at the microstructural level. Hysteresivity is independent of frequency and volume. Typical values range from 0.1 to 0.3 [Fredberg and Stamenovic, 1989].

7.6 Respiratory Control

Control of respiration occurs in many different cerebral structures [Johnson, 1991] and regulates many things [Hornbein, 1981]. Respiration must be controlled to produce the respiratory rhythm, ensure adequate gas exchange, protect against inhalation of poisonous substances, assist in maintenance of body pH, remove irritations, and minimize energy cost. Respiratory control is more complex than cardiac control for at least three reasons:

1. Airways airflow occurs in both directions.
2. The respiratory system interfaces directly with the environment outside the body.
3. Parts of the respiratory system are used for other functions, such as swallowing and speaking.

As a result, respiratory muscular action must be exquisitely coordinated; it must be prepared to protect itself against environmental onslaught, and breathing must be temporarily suspended on demand.

All control systems require sensors, controllers, and effectors. Figure 7.4 presents the general scheme for respiratory control. There are mechanoreceptors throughout the respiratory system. For example, nasal receptors are important in sneezing, apnea (cessation of breathing), bronchodilation, bronchoconstriction, and the secretion of mucus. Laryngeal receptors are important in coughing, apnea, swallowing, bronchoconstriction, airway mucus secretion, and laryngeal constriction. Tracheobronchial receptors are important in coughing, pulmonary hypertension, bronchoconstriction, laryngeal constriction, and mucus production. Other mechanoreceptors are important in the generation of the respiratory pattern and are involved with respiratory sensation.

Respiratory chemoreceptors exist peripherally in the aortic arch and carotic bodies and centrally in the ventral medulla oblongata of the brain. These receptors are sensitive to partial pressures of CO_2 and O_2 and to blood pH.

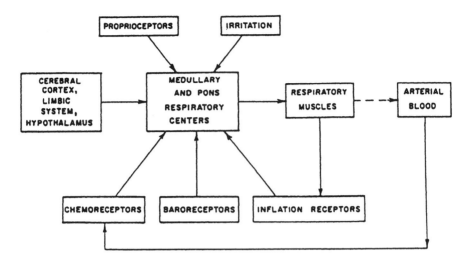

FIGURE 7.4 General scheme of respiratory control.

The respiratory controller is located in several places in the brain. Each location appears to have its own function. Unlike the heart, the basic respiratory rhythm is not generated within the lungs but rather in the brain and is transmitted to the respiratory muscles by the phrenic nerve.

Effector organs are mainly the respiratory muscles, as described previously. Other effectors are muscles located in the airways and tissues for mucus secretion. Control of respiration appears to be based on two criteria: (1) removal of excess CO_2 and (2) minimization of energy expenditure. It is not the lack of oxygen that stimulates respiration but increased CO_2 partial pressure that acts as a powerful respiratory stimulus. Because of the buffering action of blood bicarbonate, blood pH usually falls as more CO_2 is produced in the working muscles. Lower blood pH also stimulates respiration.

A number of respiratory adjustments are made to reduce energy expenditure during exercise: Respiration rate increases, the ratio of inhalation time to exhalation time decreases, respiratory flow waveshapes become more trapezoidal, and expiratory reserve volume decreases. Other adjustments to reduce energy expenditure have been theorized but not proven [Johnson, 1991].

7.7 The Pulmonary Function Laboratory

The purpose of a pulmonary function laboratory is to obtain clinically useful data from patients with respiratory dysfunction. The pulmonary function tests (PFTs) within this laboratory fulfill a variety of functions. They permit (1) quantification of a patient's breathing deficiency, (2) diagnosis of different types of pulmonary diseases, (3) evaluation of a patient's response to therapy, and (4) preoperative screening to determine whether the presence of lung disease increases the risk of surgery.

Although PFTs can provide important information about a patient's condition, the limitations of these tests must be considered. First, they are nonspecific in that they cannot determine which portion of the lungs is diseased, only that the disease is present. Second, PFTs must be considered along with the medical history, physical examination, x-ray examination, and other diagnostic procedures to permit a complete evaluation. Finally, the major drawback to *some PFTs* is that they require full patient cooperation and for this reason cannot be conducted on critically ill patients. Consider some of the most widely used PFTs: spirometry, body plethysmography, and diffusing capacity.

Spirometry

The simplest PFT is the spirometry maneuver. In this test, the patient inhales to total lung capacity (TLC) and exhales forcefully to residual volume. The patient exhales into a displacement bell chamber that sits

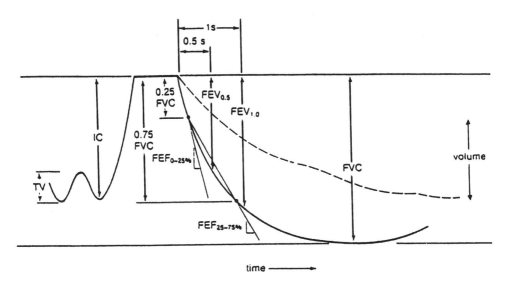

FIGURE 7.5 Typical spirometry tracing obtained during testing; inspiratory capacity (IC), tidal volume (TV), forced vital capacity (FVC), forced expiratory volume (FEV), and forced expiratory flows. Dashed line represents a patient with obstructive lung disease; solid line represents a normal, healthy individual.

on a water seal. As the bell rises, a pen coupled to the bell chamber inscribes a tracing on a rotating drum. The spirometer offers very little resistance to breathing; therefore, the shape of the spirometry curve (Fig. 7.5) is purely a function of the patient's lung compliance, chest compliance, and airway resistance. At high lung volumes, a rise in intrapleural pressure results in greater expiratory flows. However, at intermediate and low lung volumes, the expiratory flow is independent of effort after a certain intrapleural pressure is reached.

Measurements made from the spirometry curve can determine the degree of a patient's ventilatory obstruction. Forced vital capacity (FVC), forced expiratory volumes (FEV), and forced expiratory flows (FEF) can be determined. The FEV indicates the volume that has been exhaled from TLC for a particular

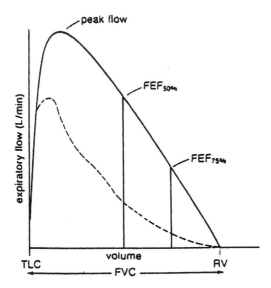

FIGURE 7.6 Flow-volume curve obtained from a spirometry maneuver. Solid line is a normal curve; dashed line represents a patient with obstructive lung disease.

time interval. For example, $FEV_{0.5}$ is the volume exhaled during the first half-second of expiration, and $FEV_{1.0}$ is the volume exhaled during the first second of expiration; these are graphically represented in Fig. 7.5. Note that the more severe the ventilatory obstruction, the lower are the timed volumes ($FEV_{0.5}$ and $FEV_{1.0}$). The FEF is a measure of the average flow (volume/time) over specified portions of the spirometry curve and is represented by the slope of a straight line drawn between volume levels. The average flow over the first quarter of the forced expiration is the $FEF_{0-25\%}$, whereas the average flow over the middle 50% of the FVC is the $FEF_{25-75\%}$. These values are obtained directly from the spirometry curves. The less steep curves of obstructed patients would result in lower values of $FEF_{0-25\%}$ and $FEF_{25-75\%}$ compared with normal values, which are predicted on the basis of the patient's sex, age, and height. Equations for normal values are available from statistical analysis of data obtained from a normal population. Test results are then interpreted as a percentage of normal.

Another way of presenting a spirometry curve is as a flow-volume curve. Figure 7.6 represents a typical flow-volume curve. The expiratory flow is plotted against the exhaled volume, indicating the maximum flow that may be reached at each degree of lung inflation. Since there is no time axis, a time must mark the $FEV_{0.5}$ and $FEV_{1.0}$ on the tracing. To obtain these flow-volume curves in the laboratory, the patient usually exhales through a *pneumotach*. The most widely used pneumotach measures a pressure drop across a flow-resistive element. The resistance to flow is constant over the measuring range of the device; therefore, the pressure drop is proportional to the flow through the tube. This signal, which is indicative of flow, is then integrated to determine the volume of gas that has passed through the tube.

Another type of pneumotach is the heated-element type. In this device, a small heated mass responds to airflow by cooling. As the element cools, a greater current is necessary to maintain a constant temperature. This current is proportional to the airflow through the tube. Again, to determine the volume that has passed through the tube, the flow signal is integrated.

The flow-volume loop in Fig. 7.7 is a dramatic representation displaying inspiratory and expiratory curves for both normal breathing and maximal breathing. The result is a graphic representation of the patient's reserve capacity in relation to normal breathing. For example, the normal patient's tidal breathing loop is small compared with the patient's maximum breathing loop. During these times of stress, this tidal breathing loop can be increased to the boundaries of the outer ventilatory loop. This increase in ventilation provides the greater gas exchange needed during the stressful situation. Compare this condition with that of the patient with obstructive lung disease. Not only is the tidal breathing loop larger than normal, but the maximal breathing loop is smaller than normal. The result is a decreased ventilatory reserve, limiting the individual's ability to move air in and out of the lungs. As the disease progresses, the outer loop becomes smaller, and the inner loop becomes larger.

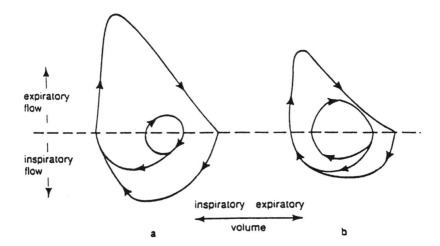

FIGURE 7.7 Typical flow-volume loops. (a) Normal flow-volume loop. (b) Flow-volume loop of patient with obstructive lung disease.

The primary use of spirometry is in detection of obstructive lung disease that results from increased resistance to flow through the airways. This can occur in several ways:

1. Deterioration of the structure of the smaller airways that results in early airways closure.
2. Decreased airway diameters caused by bronchospasm or the presence of secretions increases the airway's resistance to airflow.
3. Partial blockage of a large airway by a tumor decreases airway diameter and causes turbulent flow.

Spirometry has its limitations, however. It can measure only ventilated volumes. It cannot measure lung capacities that contain the residual volume. Measurements of TLC, FRC, and RV have diagnostic value in defining lung overdistension or restrictive pulmonary disease; the body plethysmograph can determine these absolute lung volumes.

Body Plethysmography

In a typical plethysmograph, the patient is put in an airtight enclosure and breathes through a pneumotach. The flow signal through the pneumotach is integrated and recorded as tidal breathing. At the end of a normal expiration (at FRC), an electronically operated shutter occludes the tube through which the patient is breathing. At this time the patient pants lightly against the occluded airway. Since there is no flow, pressure measured at the mouth must equal alveolar pressure. But movements of the chest that compress gas in the lung simultaneously rarify the air in the plethysmograph, and vice versa. The pressure change in the plethysmograph can be used to calculate the volume change in the plethysmograph, which is the same as the volume change in the chest. This leads directly to determination of FRC.

At the same time, alveolar pressure can be correlated to plethysmographic pressure. Therefore, when the shutter is again opened and flow rate is measured, airway resistance can be obtained as the ratio of alveolar pressure (obtainable from plethysmographic pressure) to flow rate [Carr and Brown, 1993]. Airway resistance is usually measured during panting, at a nominal lung volume of FRC and flow rate of ±1 liter/s.

Airway resistance during inspiration is increased in patients with asthma, bronchitis, and upper respiratory tract infections. Expiratory resistance is elevated in patients with emphysema, since the causes of increased expiratory airway resistance are decreased driving pressures and the airway collapse. Airway resistance also may be used to determine the response of obstructed patients to bronchodilator medications.

Diffusing Capacity

So far the mechanical components of airflow through the lungs have been discussed. Another important parameter is the diffusing capacity of the lung, the rate at which oxygen or carbon dioxide travel from the alveoli to the blood (or vice versa for carbon dioxide) in the pulmonary capillaries. Diffusion of gas across a barrier is directly related to the surface area of the barrier and inversely related to the thickness. Also, diffusion is directly proportional to the solubility of the gas in the barrier material and inversely related to the molecular weight of the gas.

Lung diffusing capacity (D_L) is usually determined for carbon monoxide but can be related to oxygen diffusion. The popular method of measuring carbon monoxide diffusion utilizes a rebreathing technique in which the patient rebreathes rapidly in and out of a bag for approximately 30 s. Figure 7.8 illustrates the test apparatus. The patient begins breathing from a bag containing a known volume of gas consisting of 0.3% to 0.5 carbon monoxide made with heavy oxygen, 0.3% to 0.5% acetylene, 5% helium, 21% oxygen, and a balance of nitrogen. As the patient rebreathes the gas mixture in the bag, a modified mass spectrometer continuously analyzes it during both inspiration and expiration. During this rebreathing procedure, the carbon monoxide disappears from the patient-bag system; the rate at which this occurs is a function of the lung diffusing capacity.

The helium is inert and insoluble in lung tissue and blood and equilibrates quickly in unobstructed patients, indicating the dilution level of the test gas. Acetylene, on the other hand, is soluble in blood and is used to determine the blood flow through the pulmonary capillaries. Carbon monoxide is bound

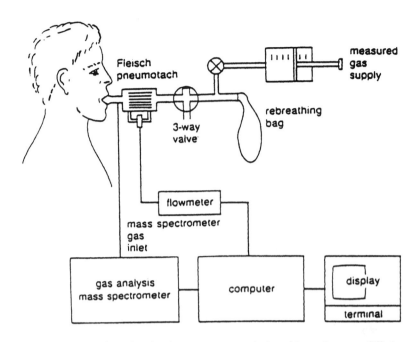

FIGURE 7.8   Typical system configuration for the measurement of rebreathing pulmonary diffusing capacity.

very tightly to hemoglobin and is used to obtain diffusing capacity at a constant pressure gradient across the alveolar-capillary membrane.

Decreased lung diffusing capacity can occur from the thickening of the alveolar membrane or the capillary membrane as well as the presence of interstitial fluid from edema. All these abnormalities increase the barrier thickness and cause a decrease in diffusing capacity. In addition, a characteristic of specific lung diseases is impaired lung diffusing capacity. For example, fibrotic lung tissue exhibits a decreased permeability to gas transfer, whereas pulmonary emphysema results in the loss of diffusion surface area.

Defining Terms

Alveoli: Respiratory airway terminals where most gas exchange with the pulmonary circulation takes place.

Diffusion: The process whereby a material moves from a region of higher concentration to a region of lower concentration.

BTPS: Body temperature (37°C) and standard pressure (1 atm), saturated ($6.28 \ \text{kN/m}^2$).

Chemoreceptors: Neural receptors sensitive to chemicals such as gas partial pressures.

Dead space: The portion of the respiratory system that does not take part in gas exchange with the blood.

Expiration: The breathing process whereby air is expelled from the mouth and nose. Also called *exhalation.*

Functional residual capacity: The lung volume at rest without breathing.

Inspiration: The breathing process whereby air is taken into the mouth and noise. Also called *inhalation.*

Mass spectrometer: A device that identifies relative concentrations of gases by means of mass-to-charge ratios of gas ions.

Mechanoreceptors: Neural receptors sensitive to mechanical inputs such as stretch, pressure, irritants, etc.

Partial pressure: The pressure that a gas would exert if it were the only constituent.

Perfusion: Blood flow to the lungs.

Plethysmography: Any measuring technique that depends on a volume change.

Pleura: The membrane surrounding the lung.

Pneumotach: A measuring device for airflow.

Pulmonary circulation: Blood flow from the right cardiac ventricle that perfuses the lung and is in intimate contact with alveolar membranes for effective gas exchange.

STPD: Standard temperature (0°C) and pressure (1 atm), dry (moisture removed).

Ventilation: Airflow to the lungs.

References

Astrand PO, Rodahl K. 1970. *Textbook of Work Physiology*. New York, McGraw-Hill.

Carr JJ, Brown JM, 1993. *Introduction to Biomedical Equipment Technology*. Englewood Cliffs, NJ, Prentice-Hall.

Fredberg JJ, Stamenovic D. 1989. On the imperfect elasticity of lung tissue. *J Appl Physiol* 67(6):2408–2419.

Hildebrandt J. 1970. Pressure-volume data of cat lung interpreted by plastoelastic, linear viscoelastic model. *J Appl Physiol* 28(3):365–372.

Hornbein TF (ed). 1981. *Regulation of Breathing*. New York, Marcel Dekker.

Johnson AT. 1991. *Biomechanics and Exercise Physiology*. New York, Wiley.

Jones NL. 1984. Normal values for pulmonary gas exchange during exercise. *Am Rev Respir Dis* 129:544–546.

Kline J. (ed). 1976. *Biologic Foundations of Biomedical Engineering*. Boston, Little, Brown.

Parker JF Jr, West VR. (eds). 1973. *Bioastronautics Data Book*. Washington, NASA.

Ruch TC, Patton HD. (eds). 1966. *Physiology Biophysics*. Philadelphia, Saunders.

Stamenovic D, Glass GM, Barnas GM, Fredberg JJ. 1990. Viscoplasticity of respiratory tissues. *J Appl Physiol* 69(3):973–988.

Suki B, Bates JHT. 1991. A nonlinear viscoelastic model of lung tissue mechanics. *J Appl Physiol* 71(3):826–833.

Weibel, ER. 1963. *Morphometry of the Human Lung*. New York, Academic Press.

West J. 1962. Regional differences in gas exchange in the lung of erect man. *J Appl Physiol* 17:893–898.

West JB (ed). 1977. *Bioengineering Aspects of the Lung*. New York, Marcel Dekker.

Additional References

Fredberg JJ, Jones KA, Nathan A, Raboudi S, Prakash YS, Shore SA, Butler JP, Sieck GC. 1996. Friction in airway smooth muscle: mechanism, latch, and implications in asthma. *J Appl Physiol* 81(6):2703–2712.

Hantos Z, Daroczy B, Csendes T, Suki B, Nagy S. 1990. Modeling of low-frequency pulmonary impedance in dogs. *J Appl Physiol* 68(3):849–860.

Hantos Z, Daroczy B, Suki B, Nagy S. 1990. Low-frequency respiratory mechanical impedance in rats. *J Appl Physiol* 63(1):36–43.

Hantos Z, Petak F, Adamicza A, Asztalos T, Tolnai J, Fredberg JJ. 1997. Mechanical impedance of the lung periphery. *J Appl Physiol* 83(5):1595–1601.

Maksym GN, Bates JHT. 1997. A distributed nonlinear model of lung tissue elasticity. *J Appl Physiol* 82(1):32–41.

Petak F, Hall GL, Sly PD. 1998. Repeated measurements of airway and parenchymal mechanics in rats by using low frequency oscillations. *J Appl Physiol* 84(5):1680–1686.

Thorpe CW, Bates JHT. 1997. Effect of stochastic heterogeneity on lung impedance during acute bronchoconstriction: A model analysis. *J Appl Physiol* 82(5):1616–1625.

Yuan H, Ingenito EP, Suki B. 1997. Dynamic properties of lung parenchyma: mechanical contributions of fiber network and interstitial cells. *J Appl Physiol* 83(5):1420–1431.

II

Biotechnology

Martin L. Yarmush and Mehmet Toner
Massachusetts General Hospital, Harvard Medical School,
and the Shriners Burns Hospital

T HE TERM *BIOTECHNOLOGY* HAS UNDERGONE significant change over the past 50 years or so. During the period prior to the 1980s, biotechnology referred primarily to the use of microorganisms for large-scale industrial processes such as antibiotic production. Since the 1980s, with the advent of recombinant DNA technology, monoclonal antibody technology, and new technologies for studying and handling cells and tissues, the field of biotechnology has undergone a tremendous resurgence in a wide range of applications pertinent to industry, medicine, and science in general. It is some of these new ideas, concepts, and technologies that will be covered in this section. We have assembled a set of chapters that covers most topics in biotechnology that might interest the practicing biomedical engineer. Absent by design is coverage of agricultural, bioprocess, and environmental biotechnology, which is beyond the scope of this handbook.

Chapter 8 deals with our present ability to manipulate genetic material. This capability, which provides the practitioner with the potential to generate new proteins with improved biochemical and physico-chemical properties, has led to the formation of the field of protein engineering. Chapter 9 discusses the field of monoclonal antibody production in terms of its basic technology, diverse applications, and ways that the field of recombinant DNA technology is currently "reshaping" some of the earlier constructs. Chapters 10 and 11 describe applications of nucleic acid chemistry. The burgeoning field of antisense technology is introduced with emphasis on basic techniques and potential applications to AIDS and cancer, and Chapter 11 is dedicated toward identifying the computational, chemical, and machine tools which are being developed and refined for genome analysis. Applied virology is the implied heading for Chapters 12 and 13, in which viral vaccines and viral-mediated gene therapy are the main foci.

Finally, Chapters 14 and 15 focus on important aspects of cell structure and function. These topics share a common approach toward quantitative analysis of cell behavior in order to develop the principles for cell growth and function. By viewing the world of biomedical biotechnology through our paradigm of proteins and nucleic acids to viruses to cells, today's biomedical engineer will hopefully be prepared to meet the challenge of participating in the greater field of biotechnology as an educated observer at the very least.

8

Protein Engineering

Alan J. Russell
University of Pittsburgh

Chenzhao Vierheller
University of Pittsburgh

Enzymes have numerous applications in both research and industry. The conformation of proteins must be maintained in order for them to function at optimal activity. Protein stability is dependent on maintaining a balance of forces that include hydrophobic interactions, hydrogen bonding, and electrostatic interactions. Most proteins are denatured, i.e., lose their active conformation in high-temperature environments (exceptions to this are found among the enzymes from thermophilic microorganisms). Therefore, understanding and maintaining enzyme stability are critical if enzymes are to be widely used in medicine and industry. Protein engineering [1] is used to construct and analyze modified proteins using molecular biologic, genetic engineering, biochemical, and traditional chemical methods (Table 8.1). The generation of proteins with improved activity and stability is now feasible. Recent developments in molecular biology also have enabled a rapid development of the technologies associated with protein engineering (Table 8.2).

8.1   Site-Direction Mutagenesis

Since site-directed mutagenesis was described by Hutchinson et al. [2] in 1978, it has become a powerful tool to study the molecular structure and function of proteins. The purpose of site-directed mutagenesis is to alter a recombinant protein by introducing, replacing, or deleting a specific amino acid. The technique enables a desired modification to be achieved with exquisite precision [3]. Site-directed mutagenesis has been used to change the activity and stability of enzymes, as well as substrate specificity and affinity. Indeed, much of the biologic detergent enzyme sold in the United States (~$200 million per year) is a protein-engineered variant of the native enzyme.

The basic idea behind site-directed mutagenesis is illustrated in Fig. 8.1. The first step of site-directed mutagenesis involves cloning the gene for the protein to be studied into a vector. This nontrivial exercise is discussed elsewhere in this handbook. Next, an oligonucleotide is designed and synthesized. The oligonucleotide will have a centrally located desired mutation (usually a mismatch) that is flanked by sequences of DNA that are complementary to a specific region of interest. Thus the oligonucleotide is designed to bind to a single region of the target gene. The mutation can then be introduced into the gene by hydridizing the oligonucleotide to the single-stranded template. Single-stranded templates of a target gene may be obtained by either cloning the gene to a single-stranded vector, such as bacteriophage M13 [4], or by using phagemids (a chimeric plasmid containing a filamentous bacteriophage replication origin that directs synthesis of single-stranded DNA with a helper bacteriophage) [5]. Alternatively, single-stranded template may be generated by digesting double-stranded DNA with exonuclease III following nicking of the target DNA with DNase or a restriction endonuclease [6].

TABLE 8.1   Selected Methods for Modification of Proteins

Chemical/biochemical methods	Side chain/amino acid residue modification
	Immobilization
Biologic methods	Site-directed mutagenesis
	Random mutagenesis

The second strand is synthesized with DNA polymerase using the oligonucleotide as a primer, and the DNA can be circularized with DNA ligase. The vector that carries the newly synthesized DNA (in which one strand is mutated) can be introduced into a bacterial host, where, because DNA duplicates in a semiconservative mode, half the newly synthesized cells containing the DNA will theoretically contain the mutation. In reality, premature DNA polymerization, DNA mismatch repair, and strand displacement synthesis result in much lower yield of mutants. Different strategies have been developed to obtain a higher level of mutation efficiency in order to minimize the screening of mutant. A second mutagenic oligonucleotide containing an active antibiotic resistance gene can be introduced at the same time as the site-directed oligonucleotide. Therefore, the wild-type plasmid will be eliminated in an antibiotic-selective medium [7] (Fig. 8.2). Alternatively, introducing a second mutagenic oligonucleotide can result in the elimination of a unique restriction endonuclease in the mutant strand [8]. Therefore, wild-type plasmid can be linearized with this unique restriction endonuclease, while the circular mutant plasmid can be transformed into bacteria host. Another alternative method is to perform second-strand extension in the presence of 5-methyl-dCTP, resulting in resistance of number of restriction enzymes, including HpaII, MspI, and Sau3A I, in the mutant strand only [9]. The template DNA can then be nicked with these enzymes, followed by digestion with exonuclease III, to increase mutagenesis efficiency.

The recent development of the polymerase chain reaction provides a new approach to site-direct mutagenesis. In 1989, Ho and colleagues [10] developed a method named *overlap extension* (Fig. 8.3), in which four oligonucleotides are used as primers. Two of the primers containing the mutant are complementary to each other. The two other primers are complementary to the opposite strand of the ends of the cloned genes. Polymerase chain reactions are performed three times. The first two polymerase chain reactions are carried out using one end primer and one mutant primer in each reaction. The products contain one double-stranded DNA from one end to the mutation point and the other double-stranded DNA from another end to the mutation point. In other words, the two DNA products are overlapped in the mutated region. The polymerase chain reaction products are purified and used as templates for the third polymerase chain reaction with two end primers. This generates the whole length of DNA with the desired mutation. The advantage of this method is that it can be done quickly with nearly 100% efficiency.

After screening and selection, a mutation is generally confirmed by DNA sequencing. The mutant can then be subcloned into an expression vector to test the effect of mutation on the activity of the enzyme.

TABLE 8.2   Potentially Useful Modifications to Proteins

Stability	Increased thermostability
	Increased stability at extremes of pH
	Increased stability in organic solvents
	Resistance to oxidative inactivation
	Resistance to proteolysis
Kinetics	Increased maximum velocity
	Altered affinity for substrate
	Altered substrate specificity
	Resistance to substrate/product inhibition
Biology	Altered spectrum of activity
	Altered substrate specificity

Source: Modified from Primrose SB. 1991. In *Molecular Biotechnology*, 2d ed. New York, Blackwell Scientific.

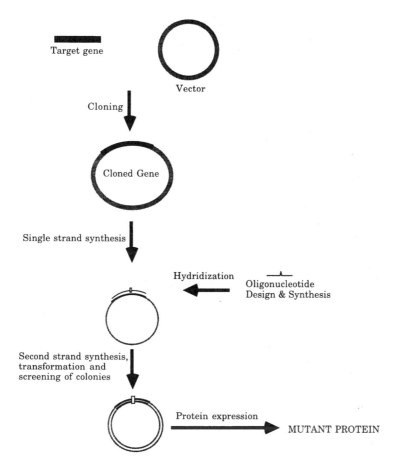

FIGURE 8.1 Overview of site-directed mutagenesis. See text for details.

Over the last decade, site-directed mutagenesis has become a somewhat trivial technical process, and the most challenging segment of a protein engineering research project is unquestionably the process of determining why a mutant protein has altered properties. The discussion of this particular enterprise lies beyond the scope of this brief review.

8.2 Solvent Engineering

An enzyme-catalyzed reaction can be simplified to its most basic components by considering it as the transfer of a substrate molecule from solvent to the surface of an enzyme molecule. The exchange of substrate-solvent and enzyme-solvent interactions for enzyme-substrate interactions then enables the chemistry of catalysis to take place. *Protein engineering,* as described above, is the process of changing the enzyme in a predictable and precise manner to effect a change on the catalytic process. Since the enzyme is only one side of the balance, however, any changes in the rest of the equation also will alter the catalytic process. Until the late 1980s, for example, substrate specificity could be altered by either protein engineering of the enzyme or by changing the substrate. *Solvent engineering* is now also emerging as a powerful tool in rational control of enzyme activity. Using solvents other than water has successfully led to enzymes with increased thermostability, activity against some substrates, pH dependence, and substrate and enantiospecificity.

The simplicity of the solvent-engineering approach is clear. If a protein is not inactivated by a solvent change, then its activity will be dependent on the solvent in which the enzyme and substrate are placed.

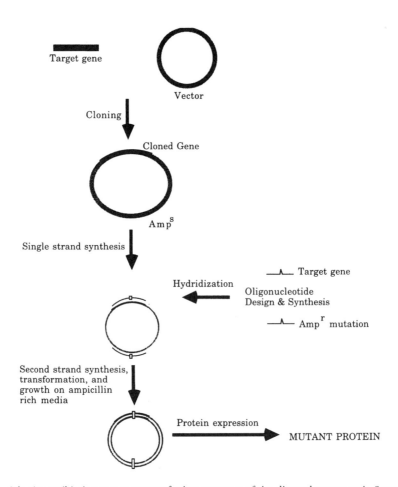

FIGURE 8.2 Selective antibiotic genes as strategy for improvement of site-directed mutagenesis. See text for details.

This strategy has been discussed in detail in recent reviews [11], and we will only summarize the most pertinent information here.

Enzymes that have been freeze-dried and then suspended in anhydrous organic solvents, in which proteins are not soluble, retain their activity and specificity [12]. The enzyme powders retain approximately a monolayer of water per molecule of enzyme during the freeze-drying process. As long as this monolayer of water remains associated with the enzyme in an organic environment, the structure of the enzyme is not disrupted [13], and hence enzyme activity in essentially anhydrous organic solvents can be observed.

The physical properties of a solvent in which an enzyme is placed can influence the level of activity and specificity in a given reaction. For instance, alcohol dehydrogenase can catalyze oxidation-reduction reactions equally well in buffer and heptane, but the level of activity in heptane is sharply dependent on the thermodynamic activity of water in the system. In general, activity can be increased by increasing the hydrophobicity of the solvent used. Interestingly, however, the specificity (both stereo- and substrate) of an enzyme or antibody dispersed in an organic solvent is generally greater in more hydrophilic solvents [14]. Water is involved as a substrate in many of the chemical modification processes that lead to irreversible inaction of proteins. Not surprisingly, therefore, proteins in anhydrous environments are stable at temperatures exceeding 100°C [15].

Considerable attention is now being given to developing the predictability of the solvent-engineering approach. Elucidating the structure-function-environment relationships that govern the activity and specificity of an enzyme is a crucial step for learning how to apply the power of solvent engineering. In recent

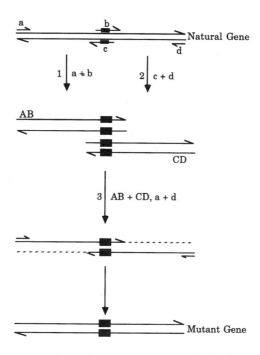

FIGURE 8.3 Site-directed mutagenesis by overlap extension. See text for details.

years, supercritical fluids (materials at temperatures and pressures above their critical point) have been used as a dispersant for enzyme-catalyzed reactions. In addition to being excellent process solvents, the physical properties of supercritical fluids are pressure-dependent. Thus it may be possible to detect which physical properties of a nonaqueous medium have a role in determining a given enzyme function [16].

8.3   Conclusions

The goals of protein and solvent engineering are similar. Ideally, one should be able to predictably alter a given property of an enzyme by either changing the enzyme or its environment. In reality, both approaches are still somewhat unpredictable. Both protein engineering and solvent engineering have been used successfully to alter the protein properties that appear in Table 8.1. Further research is needed, however, before the results of any given attempt at such biologic engineering can be predicted.

References

1. Alvaro G, Russell AJ. 1991. *Methods Enzymol* 202:620.
2. Hutchinson C, Phillips S, Edgell M. 1978. *J Biol Chem* 253:6551.
3. Moody P, Wilkinson A. 1990. *Protein Engineering*, pp 1–3. IRL Press.
4. Zoller M, Smith M. 1983. *Methods Enzymol* 100:468.
5. Dente L, Cesareni G, Cortese R. 1983. *Nucl Acid Res* 11:1645.
6. Rossi J, Zoller M. 1987. In D Oxender, C Fox (eds), *Protein Engineering*, pp 51–63. New York, Alan R Liss.
7. *Promega Protocols and Application Guide*, 2d ed, pp 98–105. 1991.
8. Deng W, Nickoloff J. 1992. *Anal Biochem* 200:81.
9. Vandeyar M, Weiner M, Hutton C, Batt C. 1988. *Gene* 65:129.
10. Ho N, Hunt H, Horton R, et al. 1989. *Gene* 77:51.
11. Russell AJ, Chatterjee S, Rapanovich I, Goodwin J. 1992. In A Gomez-Puyon (ed), *Biomolecules in Organic Solvents*, pp 92–109. Boca Raton, Fla, CRC Press.

12. Zaks A, Klibanov AM. 1985. *Proc Natl Acad Sci USA* 82:3192.
13. Affleck R, Xu Z-F, Suzawa V, et al. 1992. *Proc Natl Acad Sci USA* 89:1100.
14. Kamat SV, Beckman EJ, Russell AJ. 1993. *J Am Chem Soc* 115:8845.
15. Zaks A, Klibanov AM. 1984. *Science* 224:1249.
16. Kamat S, Iwaskewycz B, Beckman EJ, Russell AJ. 1993. *Proc Natl Acad Sci USA* 90:2940.

Monoclonal Antibodies and Their Engineered Fragments

Srikanth Sundaram
Rutgers University

David M. Yarmush
Rutgers University

Antibodies are a class of topographically homologous multidomain glycoproteins produced by the immune system that display a remarkably diverse range of binding specificities. The most important aspects of the immune system are that it is diverse and driven to produce antibodies of the highest possible antigen affinity. The primary repertoire of antibodies consists of about 10^9 different specificities, each of which can be produced by an encounter with the appropriate antigen. This diversity is known to be produced by a series of genetic events each of which can play a role in determining the final function of the antibody molecule. After the initial exposure to the antigen, additional diversity occurs by a process of somatic mutation so that, for any selected antigen, about 10^4 new binding specificities are generated. Thus the immunologic repertoire is the most diverse system of binding proteins in biology. Antibodies also display remarkable binding specificity. For example, it has been shown that antibodies are able to distinguish between *ortho-*, *meta-*, and *para*-forms of the same haptenic group [Landsteiner, 1945]. This exquisite specificity and diversity make antibodies ideal candidates for diagnostic and therapeutic agents.

Originally, the source of antibodies was antisera, which by their nature are limited in quantity and heterogeneous in quality. Antibodies derived from such sera are termed *conventional antibodies* (*polyclonal antibodies*). Polyclonal antibody production requires methods for the introduction of immunogen into animals, withdrawal of blood for testing the antibody levels, and finally exsanguination for collection of immune sera. These apparently simple technical requirements are complicated by the necessity of choosing a suitable species and immunization protocol that will produce a highly immune animal in a short time. Choice of animal is determined by animal house facilities available, amount of antiserum required (a mouse will afford only 1.0 to 1.5 ml of blood; a goat can provide several liters), and amount of immunogen available (mice will usually respond very well to 50 µg or less of antigen; goats may require several milligrams). Another consideration is the phylogenic relationship between the animal from which the immunogen is derived and that used for antibody production. In most cases, it is advisable to immunize a species phylogenetically unrelated to the immunogen donor, and for highly conserved mammalian proteins, nonmammals (e.g., chickens) should be used for antibody production. The poly-clonal antibody elicited by an antigen facilitates the localization, phagocytosis, and complement-mediated lysis of that antigen; thus the usual polyclonal immune response has clear advantages in vivo. Unfortu-nately, the antibody heterogeneity that increases immune protection *in vivo* often reduces the efficacy of an antiserum for various *in vitro* uses. Conventional heterogeneous antisera vary from animal to animal and contain undesirable nonspecific or cross-reacting antibodies. Removal of unwanted specificities from a polyclonal antibody preparation is a time-consuming task, involving repeated adsorption techniques, which often results in the loss of much of the desired antibody and seldom is very effective in reducing the heterogeneity of an antiserum.

After the development of hybridoma technology [Köhler and Milstein, 1975], a potentially unlimited quantity of homogeneous antibodies with precisely defined specificities and affinities (*monoclonal anti-bodies*) became available, and this resulted in a step change in the utility of antibodies. Monoclonal antibodies (mAbs) have gained increasing importance as reagents in diagnostic and therapeutic medicine, in the identification and determination of antigen molecules, in biocatalysis (catalytic antibodies), and in affinity purification and labeling of antigens and cells.

9.1 Structure and Function of Antibodies

Antibody molecules are essentially required to carry out two principal roles in immune defense. First, they recognize and bind nonself or foreign material (antigen binding). In molecular terms, this generally means binding to structures on the surface of the foreign material (antigenic determinants) that differ from those on the host. Second, they trigger the elimination of foreign material (biologic effector functions). In molecular terms, this involves binding of effector molecules (such as complement) to the antibody-antigen complex to trigger elimination mechanisms such as the complement system and phago-cytosis by macrophages and neutrophils, etc.

In humans and other animals, five major immunoglobulin classes have been identified. The five classes include immunoglobulins G (IgG), A (IgA), M (IgM), D (IgD), and E (IgE). With the exception of IgA (dimer) and IgM (pentamer), all other antibody classes are monomeric. The monomeric antibody molecule consists of a basic four-chain structure, as shown in Fig. 9.1. There are two distinct types of chains: the light (L) and the heavy chains (H). The chains are held together by disulfide bonds. The light and heavy chains are held together by interchain disulfides, and the two heavy chains are held together by numerous disulfides in the hinge region of the heavy chain. The light chains have a molecular weight of about 25,000 Da, while the heavy chains have a molecular weight of 50,000 to 77,000 Da depending on the isotope. The L chains can be divided into two subclasses, kappa (κ) and lambda (λ), on the basis of their structures and amino acid sequences. In humans, about 65% of the antibody molecules have κ chains, whereas in rodents, they constitute over 95% of all antibody molecules. The light chain consists of two structural domains: The carboxy-terminal half of the chain is constant except for

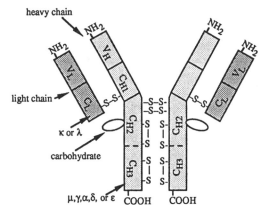

FIGURE 9.1 The structure of a monomeric antibody molecule.

certain allotypic and isotype variations and is called the C_L (constant: light chain) region, whereas the amino-terminal half of the chain shows much sequence variability and is known as the V_L (variable:

light chain) region. The H chains are unique for each immunoglobulin class and are designated by the Greek letter corresponding to the capital letter designation of the immunoglobulin class (α for the H chains of IgA, γ for the H chains of IgG). IgM and IgA have a third chain component, the J chain, which joins the monomeric units. The heavy chain usually consists of four domains: The amino-terminal region (approximately 110 amino acid residues) shows high sequence variability and is called the V_H (variable: heavy chain) region, whereas there are three domains called C_{H1}, C_{H2}, and C_{H3} in the constant part of the chain. The C_{H2} domain is glycosylated, and it has been shown that glycosylation is important in some of the effector functions of antibody molecules. The extent of glycosylation varies with the antibody class and, to some extent, the method of its production. In some antibodies (IgE and IgM), there is an additional domain in the constant part of the heavy chain called the C_{H4} region. The hinge region of the heavy chain (found between the C_{H1} and C_{H2} domains) contains a large number of hydrophilic and proline residues and is responsible for the segmental flexibility of the antibody molecule.

All binding interactions to the antigen occur within the variable domains (V_H and V_L). Each variable domain consists of three regions of relatively greater variability, which have been termed *hypervariable* (HV) or *complementarity-determining regions* (CDR) because they are the regions that determine complementarity to the particular antigen. Each CDR is relatively short, consisting of from 6 to 15 residues. In general, the topography of the six combined CDRs (three from each variable domain) produces a structure that is designed to accommodate the particular antigen. The CDRs on the light chain include residues 24 to 34 (L1), 50 to 56 (L2), and 89 to 73 (L3), whereas those on the heavy chain include residues 31 to 35 (H1), 50 to 65 (H2), and 95 to 102 (H3) in the Kabat numbering system. The regions between the hypervariable regions are called the *framework regions* (FR) because they are more conserved and play an important structural role. The effector functions are, on the other hand, mediated via the two C-terminal constant domains, namely, the C_{H2} and C_{H3}.

Each variable region is composed of several different genetic elements that are separate on the germ line chromosome but rearrange in the B cell to produce a variable-region exon. The light chain is composed of two genetic elements, a variable (V) gene and a joining (J) gene. The V_κ gene generally encodes residues 1 to 95 in a kappa light chain, and the J_κ gene encodes residues 96 to 107, which is the carboxyl end of the variable region. Thus the kappa V gene encodes all the first and second hypervariable regions (L1 and L2) and a major portion of the third hypervariable region (L3). In the human system, there are estimated to be about 100 different V_κ genes and about 5 functional J_κ genes, and the potential diversity is increased by the imprecision of the VJ joining. The heavy-chain variable region is similarly produced by the splicing together of genetic elements that are distant from each other in the germ line. The mature V_H region is produced by the splicing of a variable gene segment (V) to a diversity segment (D) and to a joining (J) segment. There are about 300 V_H genes, about 5 D_H genes, and 9 functional J_H genes [Tizard, 1992]. With respect to the CDRs of the heavy chain, H1 and H2 are encoded by the V_H gene, while the H3 region is formed by the joining of the V, D, and J genes. Additional diversity is generated by recombinational inaccuracies, light- and heavy-chain recombination, and N-region additions.

The three-dimensional structure of several monoclonal antibodies has been determined via x-ray crystallography and has been reviewed extensively [Davies et al., 1990]. Each domain consists of a sandwich of β sheets with a layer of four and three strands making up the constant domains and a layer of four and five strands making up the variable domains. The CDRs are highly solvent exposed and are the loops at the ends of the variable region β strands that interact directly with the antigen. The superimposition of several V_L and V_H structures from available x-ray crystallographic coordinates using the conserved framework residues shows that the framework residues of the antibody variable domains are spatially closely superimposable. The CDRs varied in shape and length, as well as in sequence. Similarities in the structures of the CDRs from one antibody to another suggest that they are held in a fixed relationship in space, at least within a domain. Also, the N and C terminal of the CDRs were rigidly conserved by the framework residues. Among the CDRs, L1, L2, H2, and H3 show the most conformational variability.

9.2 Monoclonal Antibody Cloning Techniques

The original, and still dominant method for cloning murine monoclonal antibodies is through the fusion of antibody-producing spleen B cells and a myeloma line [Köhler and Milstein, 1975]. More recently, "second-generation" monoclonal antibodies have been cloned without resorting to hybridoma technology via the expression of combinatorial libraries of antibody genes isolated directly from either immunized or naive murine spleens or human peripheral blood lymphocytes. In addition, display of functional antigen-binding antibody fragments (Fab, sFv, Fv) on the surface of filamentous bacteriophage has further facilitated the screening and isolation of engineered antibody fragments directly from the immunologic repertoire. Repertoire cloning and phage display technology are now readily available in the form of commercial kits [e.g., ImmunoZap and SurfZap Systems from Stratacyte, Recombinant Phage Antibody System (RPAS) from Pharmacia] with complete reagents and protocols.

Hybridoma Technology

In 1975, Kohler and Milstein [1975] showed that mouse myeloma cells could be fused to B lymphocytes from immunized mice resulting in continuously growing, specific monoclonal antibody-secreting somatic cell hybrids, or *hybridoma* cells. In the fused hybridoma, the B cell contributes the capacity to produce specific antibody, and the myeloma cell confers longevity in culture and ability to form tumors in animals.

The number of fusions obtained via hybridoma technology is small, and hence there is a need to select for fused cells. The selection protocol takes advantage of the fact that in normal cells there are two biosynthetic pathways used by cells to produce nucleotides and nucleic acids. The *de novo* pathway can be blocked using aminopterin, and the salvage pathway requires the presence of functional hypoxanthine guanine phosphoribosyl transferase (HGPRT). Thus, by choosing a myeloma cell line that is deficient in HGPRT as the "fusion partner," Kohler and Milstein devised an appropriate selection protocol. The hybrid cells are selected by using the HAT (hypoxanthine, aminopterin, and thymidine) medium in which only myeloma cells that have fused to spleen cells are able to survive (since unfused myeloma cells are HGPRT−). There is no need to select for unfused spleen cells because these die out in culture.

Briefly, experimental animals (mice or rats) are immunized by injecting them with soluble antigen or cells and an immunoadjuvant. Three of four weeks later, the animals receive a booster injection of the antigen without the adjuvant. Three to five days after this second injection, the spleens of those animals that produce the highest antibody titer to the antigen are excised. The spleen cells are then mixed with an appropriate "fusion partner," which is usually a nonsecreting hybridoma or myeloma cell (e.g., SP2/0 with 8-azaguanine selection) that has lost its ability to produce hypoxanthine guanine phosphoribosyl transferase (HGPRT−). The cell suspension is then briefly exposed to a solution of polyethylene glycol (PEG) to induce fusion by reversibly disrupting the cell membranes, and the hybrid cells are selected for in HAT media. In general, mouse myelomas should yield hybridization frequencies of about 1 to 100 clones per 10^7 lymphocytes. After between 4 days and 3 weeks, hybridoma cells are visible, and culture supernatants are screened for specific antibody secretion by a number of techniques (e.g., radioimmunoassay, ELISA, immunoblotting). Those culture wells that are positive for antibody production are then expanded and subcloned by limiting dilution or in soft agar to ensure that they are derived from a single progenitor cell and to ensure stability of antibody production. Reversion of hybridomas to nonsecreting forms can occur due to loss or rearrangement of chromosomes. The subclones are, in turn, screened for antibody secretion, and selected clones are expanded for antibody production *in vitro* using culture flasks or bioreactors (yield about 10 to 100 μm/liter) or *in vivo* as ascites in mice (1 to 10 mg/ml).

The most commonly used protocol is the polyethylene glycol (PEG) fusion technique, which, even under the most efficient conditions, results in the fusion of less than 0.5% of the spleen cells, with only about 1 in 10^5 cells forming viable hybrids. Several methods have been developed to enhance conventional hybrid formation that have led to only incremental improvements in the efficiency of the fusion process (see Neil and Urnovitz [1988] for review). Methods that enhance conventional hybridoma formation include pretreatment of myeloma cells with colcemid and/or *in vitro* antigen stimulation of the spleen cells prior to fusion, addition of DMSO or phytohemagglutinin (PHA) to PEG during fusion, and addition of insulin, growth factors, human endothelial culture supernatants (HECS), etc. to the growth medium after fusion. In addition, improvements in immunization protocols such as suppression of dominant responses, *in vitro* or intrasplenic immunization, and antigen targeting also have been developed. Suppression of dominant immune responses is used to permit the expression of antibody-producing cells with specificity for poor immunogens and is achieved using the selective ability of cyclophosphoamide to dampen the immune response to a particular antigen, followed by subsequent immunization with a similar antigen. In *in vitro* immunization, spleen cells from nonimmunized animals are incubated with small quantities of antigen in a growth medium that has been "conditioned" by the growth of thymocytes. This technique is used most commonly for the production of human hybridomas, where *in vivo* immunization is not feasible. Intrasplenic immunization involves direct injection of immunogen into the spleen and is typically used when only very small quantities of antigen are available. Other advantages include a shortened immunization time, ability to generate high-affinity monoclonals, and improved diversity in the classes of antibodies generated. Finally, in antigen targeting, the immune response is enhanced by targeting the antigen to specific cells of the immune system, for example, by coupling to anti-class II monoclonal antibodies.

There also have been several advances in fusion techniques such as electrofusion, antigen bridging, and laser fusion. In electrofusion, cells in suspension are aligned using an ac current resulting in cell-cell contact. A brief dc voltage is applied that induces fusion and has resulted in a 30- to 100-fold increase in fusion frequencies in some selected cases. Further improvement in electrofusion yields have been obtained by an antigen bridging, wherein avidin is conjugated with the antigen and the myeloma cell membranes are treated with biotin. Spleen cells expressing immunoglobulins of correct specificity bind to the antigen-avidin complex and are in turn bound by the biotinylated cell membranes of the myeloma cells. Finally, laser-induced cell fusion in combination with antigen bridging has been used to eliminate the tedious and time-consuming screening process associated with traditional hybridoma techniques. Here, rather than carry out the fusion process in "bulk," preselected B cells producing antibody of desired specificity and affinity are fused to myeloma cells by irradiating each target cell pair (viewed under a microscope) with laser pulses. Each resulting hybridoma cell is then identified, subcloned, and subsequently expanded.

Despite these improvements, many of the limitations of the hybridoma technique persist. First, it is slow and tedious and labor- and cost-intensive. Second, only a few antibody-producing hybridoma lines are created per fusion, which does not provide for an adequate survey of the immunologic repertoire. Third, as is the case with most mammalian cell lines, the actual antibody production rate is low. Fourth, it is not easy to control the class or subclass of the resulting antibodies, a characteristic that often determines their biologic activity and therefore their usefulness in therapeutic applications. Finally, the production of human monoclonal antibodies by conventional hybridoma techniques has not been very successful due to a lack of suitable human fusion partners, problems related to immunization, difficulty in obtaining human lymphocytes, etc.

Repertoire Cloning Technology

The shortcomings of the hybridoma technology and the potential for improvement of molecular properties of antibody molecules by a screening approach have led to the expression of the immunologic repertoire in *Escherichia coli* (repertoire cloning). Two developments were critical to the development of this technology. First, the identification of conserved regions at each end of the nucleotide sequences encoding the variable domains enabled the use of the polymerase chain reactions (PCR) to clone antibody Fv and Fab genes from both spleen genomic DNA and spleen messenger RNA [Orlandi et al., 1989; Sastry et al., 1989]. The amplification of a target sequence via PCR requires two primers, each annealing to one end of the target gene. In the case of immunoglobulin variable-region genes, the J segments are sufficiently conserved to enable the design of universal "downstream" primers. In addition, by comparing the aligned sequences of many variable-region genes, it was found that 5′ ends of the V_H and V_L genes are also relatively conserved so as to enable the design of universal "upstream" primers as well. These primers were then used to establish a repertoire of antibody variable-region genes. Second, the successful expression of functional antigen-binding fragments in bacteria using a periplasmic secretion strategy enabled the direct screening of libraries of cloned antibody genes for antigen binding [Better et al., 1988; Skerra and Pluckthun, 1988].

The first attempt at repertoire cloning resulted in the establishment of diverse libraries of V_H genes from spleen genomic DNA of mice immunized with either lysozyme or keyhole-limpet hemocyanin (KLH). From these libraries, V_H domains were expressed and secreted in *E. coli* [Ward et al., 1989]. Binding activities were detected against both antigens in both libraries; the first library, immunized against lysozyme, yielded 21 clones with lysozyme activity and 2 with KLH activity, while the second library, immunized against KLH, yielded 14 clones with KLH activity and 2 with lysozyme activity. Two V_H domains were characterized with affinities for lysozyme in the 20 nM range. The complete sequences of 48 V_H gene clones were determined and shown to be unique. The problems associated with this single-domain approach are (1) isolated V_H domains suffer from several drawbacks such as lower selectivity and poor solubility and (2) an important source of diversity arising from the combination of the heavy and light chains is lost.

In the so-called combinatorial approach, Huse et al. [1989] used a novel bacteriophage λ vector system (λ-ZAP technology) to express in *E. coli* a combinatorial library of Fab fragments of the murine antibody repertoire. The combinatorial expression library was constructed from spleen mRNA isolated from a mouse immunized with KLH-coupled *p*-nitrophenyl phosphonamidate (NPN) antigen in two steps. In the first step, separate heavy-chain (Fd) and light-chain (κ) libraries were constructed. These two libraries were then combined randomly, resulting in the combinatorial library in which each clone potentially coexpresses a heavy and a light chain. In this case, they obtained 25 million clones in the library, approximately 60% of which coexpressed both chains. One million of these were subsequently screened for antigen binding, resulting in approximately 100 antibody-producing, antigen-binding clones. The light- and heavy-chain libraries, when expressed individually, did not show any binding activity. In addition, the vector systems used also permitted the excision of a phagemid containing the Fab genes; when grown in *E. coli*, these permitted the production of functional Fab fragments in the periplasmic supernatants. While the study did not address the overall diversity of the library, it did establish repertoire cloning as a potential alternative to conventional hybridoma technology.

Repertoire cloning via the λ-ZAP technology (now commercially available as the ImmunoZap kit from Stratacyte) has been used to generate antibodies to influenza virus hemagglutinin (HA) starting with mRNA from immunized mice [Caton and Kaprowski, 1990]. A total of 10 antigen-binding clones was obtained by screening 125,000 clones from the combinatorial library consisting of 25 million members. Partial sequence analysis of the V_H and V_κ regions of five of the HA-positive recombinants revealed that all the HA-specific antibodies generated by repertoire cloning utilized a V_H region derived from members of a single B-cell clone in conjunction with one of two light-chain variable regions. A majority of the HA-specific antibodies exhibited a common heavy–light-chain combination that was very similar to one previously identified among HA-specific hybridoma monoclonal antibodies. The relative representation of these sequences and the overall diversity of the library also were studied via hybridization studies and sequence analysis of randomly selected clones. It was determined that a single functional V_H sequence was present at a frequency of 1 in 50, while the more commonly occurring light-chain sequence was present at a frequency of 1 in 275. This indicates that the overall diversity of the gene family representation in the library is fairly limited.

The λ-ZAP technology also has been used to produce high-affinity human monoclonal antibodies specific for tetanus toxoid [Mullinax et al., 1990; Persson et al., 1991]. The source of the mRNA in these studies was peripheral blood lymphocytes (PBLs) from human donors previously immunized with tetanus toxoid and boosted with the antigen. Mullinax et al. [1990] estimated that the frequency of positive clones in their library was about 1 in 500, and their affinity constants ranged from 10^6 to 10^9 M (Molar)$^{-1}$. However, the presence of a naturally occurring SacI site (one of the restriction enzymes used to force clone PCR-amplified light-chain genes) in the gene for human C_κ may have resulted in a reduction in the frequency of positive clones. Persson et al. [1991] constructed three different combinatorial libraries using untreated PBLs, antigen-stimulated PBLs (cells cultured in the presence of the antigen), and antigen-panned PBLs (cells that were selected for binding to the antigen). Positive clones were obtained from all three libraries with frequencies of 1 in 6000, 1 in 5000, and 1 in 4000, respectively. Apparently binding constants were estimated to be in the range of 10^7 to 10^9 M^{-1}. Sequence analysis of a limited number of clones isolated from the antigen-stimulated cell library indicated a greater diversity than that described for HA or NPN. For example, of the eight heavy-chain CDR3 sequences examined, only two pairs appeared to be clonally related. The λ-ZAP technology also has been used to rescue a functional human antirhesus D Fab from an EBV-transformed cell line [Burton, 1991].

In principle, repertoire cloning would allow for the rapid and easy identification of monoclonal antibody fragments in a form suitable for genetic manipulation. It also provides for a much better survey of the immunologic repertoire than conventional hybridoma technology. However, repertoire cloning is not without its disadvantages. First, it allows for the production of only antibody fragments. This limitation can be overcome by mounting the repertoire cloned variable domains onto constant domains that possess the desired effector functions and using transfectoma technology to express the intact immunoglobulin genes in a variety of host systems. This has been demonstrated for the case of a human Fab fragment to tetanus toxoid, where the Fab gene fragment obtained via repertoire cloning was linked to an Fc fragment gene and successfully expressed in a CHO cell line [Bender et al., 1993]. The second limitation to the use of "immunized" repertoires, which has serious implications in the applicability of this technology for the production of human monoclonal antibodies. The studies reviewed above have all used spleen cells or PBLs from immunized donors. This has resulted in relatively high frequency of positive clones, eliminating the need for extensive screening. Generating monoclonal antibodies from "naive" donors (who have not had any exposure to the antigen) would require the screening of very large libraries. Third, the actual diversity of these libraries is still unclear. The studies reported above show a wide spectrum ranging from very limited (the HA studies) to moderate (NPN) to fairly marked diversity (tetanus toxoid). Finally, the combinatorial approach is disadvantageous in that it destroys the original pairing of the heavy and light chains selected for by immunization. Strategies for overcoming some of these limitations have already been developed and are reviewed below.

Phage Display Technology

A critical aspect of the repertoire cloning approach is the ability to screen large libraries rapidly for clones that possess desired binding properties, e.g., binding affinity or specificity, catalysis, etc. This is especially the case for "naive" human repertoires, wherein the host has not been immunized against the antigen of interest for ethical and/or safety reasons. In order to facilitate screening of large libraries of antibody genes, *phage display* of functional antibody fragments has been developed, which has resulted in an enormous increase in the utility of repertoire cloning technology. In phage display technology, functional antibody fragments (such as the sFv and Fab) are expressed on the surface of filamentous bacteriophages, which facilitates the selection of specific binders (or any other property such as catalysis, etc.) from a large pool of irrelevant antibody fragments. Typically, several hundreds of millions of phage particles (in a small volume of 50 to 100 ml) can be tested for specific binders by allowing them to bind to the antigen of interest immobilized to a solid matrix, washing away the nonbinders, and eluting the binders using a suitable elution protocol.

Phage display of antibody fragments is accomplished by coupling of the antibody fragment to a coat protein of the bacteriophage. Two different coat proteins have been used for this purpose, namely, the major coat protein encoded for by gene VIII and the adsorption protein encoded for by gene III. The system based on gene VIII displays several copies of the antibody fragment (theoretically there are 2000 copies of gene VIII product per phage) and is used for the selection of low-affinity binders. The gene III product is, on the other hand, present at approximately four copies per phage particle and leads to the selection of high-affinity binders. However, since the native gene III product is required for infectivity, at least one copy on the phage has to be a native one.

The feasibility of phage display of active antibody fragments was first demonstrated by McCafferty et al. [1990] when the single-chain Fv fragment (single-chain antibody) of the anti-hen egg white lysozyme (HEL) antibody was cloned into an fd phage vector at the N-terminal region of the gene III protein. This study showed that complete active sFv domains could be displayed on the surface of bacteriophage fd and that rare phage displaying functional sFv (1 in 10^6) can be isolated. Phage that bound HEL were unable to bind to turkey egg white lysozyme, which differs from HEL by only seven residues. Similarly, active Fab fragments also have been displayed on phage surfaces using gene VIII [Kang et al., 1991]. In this method, assembly of the antibody Fab molecules in the periplasm occurs in concert with phage morphogenesis. The Fd chain of the antibody fused to the major coat protein VIII of phage M13 was coexpressed with κ chains, with both chains being delivered to the periplasmic space by the pelB leader sequence. Since the Fd chain is anchored in the membrane, the concomitant secretion of the κ chains results in the assembly of the two chains and hence the display of functional Fab on the membrane surface. Subsequent infection with helper phage resulted in phage particles that had incorporated functional Fab along their entire surface. Functionality of the incorporated Fab was confirmed by antigen-specific precipitation of phage, enzyme-linked immunoassays, and electron microscopy. The production of soluble antibody fragments from selected phages can now be accomplished without subcloning [Hoogenboom et al., 1991]. The switch from surface display to soluble antibody is mediated via the use of an amber stop codon between the antibody gene and phage gene protein. In a *supE* suppresser strain of *E. coli*, the amber codon is read as Glu and the resulting fusion protein is displayed on the surface of the phage. In nonsuppresser strains, however, the amber codon is read as a stop codon, resulting in the production of soluble antibody.

The combination of repertoire cloning technology and phage display technology was initially used to screen antibody fragments from repertoires produced from immunized animals, namely, the production of Fv fragments specific for the hapten phenyloxazolone using immunized mice [Clackson et al., 1991], human Fab fragments to tetanus toxoid [Barbas et al., 1991], human Fab fragments to gp120 using lymphocytes isolated from HIV-positive individuals [Burton et al., 1991], and human Fab fragments to hepatitis B surface antigen from vaccinated individuals [Zebedee et al., 1992]. These studies established the utility of phage display as a powerful screening system for functional antibody fragments. For example, attempts to generate human Fab fragments against gp120 using the λ-ZAP technology failed to produce any binders. Phage display, on the other hand, resulted in 33 of 40 clones selected via antigen panning possessing clear reactivity with affinity constants of the order of 10^{-8} M^{-1}. In the case of the tetanus toxoid studies, phage display was used to isolate specific clones from a library that included known tetanus toxoid clones at a frequency of 1 in 170,000.

Bypassing Immunization

The next step was the application of phage display technology to generate antibodies from unimmunized donors (naive repertoires). Marks et al. [1991] constructed two combinatorial libraries starting from peripheral blood lymphocytes of unimmunized human donors, namely, an IgM library using μ-specific PCR primers and an IgG library using γ-specific primers. The libraries were then screened using phage display and sFv fragments specific for nonself antigens such as turkey egg white lysozyme, bovine serum albumin, phenyloxazolone, and bovine thyroglobulin, as well as for self antigens such as human thyroglobulin, human tumor necrosis factor α, cell surface markers, carcinoembryonic antigen, and mucin, and human blood group antigens were isolated [Hoogenboom et al., 1992]. The binders were all isolated from the IgM library (with the exception of six clones for turkey egg white lysozyme isolated from the IgG library), and the affinities of the soluble antibody fragments were low to moderate (2×10^6 to 10^7 M^{-1}). Both these results are typical of the antibodies produced during a primary response.

The second stage of an immune response *in vivo* involves affinity maturation, in which the affinities of antibodies of the selected specificities are increased by a process of somatic mutation. Thus one method by which the affinities of antibodies generated from naive repertoires may be increased is by mimicking this process. Random mutagenesis of the clones selected from naive repertoires has been accomplished by error-prone PCR [Gram et al., 1992]. In this study, low-affinity Fab fragments (10^4 to 10^5 M^{-1}) to progesterone were initially isolated from the library, and their affinities increased 13- to 30-fold via random mutagenesis. An alternative approach to improving the affinities of antibodies obtained from naive repertoires involves the use of chain shuffling [Marks et al., 1992]. This study describes the affinity maturation of a low-affinity antiphenyloxazolone antibody ($3 \times 10^7 \ M^{-1}$). First, the existing light chain was replaced with a repertoire of *in vivo* somatically mutated light chains from unimmunized donors resulting in the isolation of a clone with a 20-fold increase in affinity. Next, shuffling of the heavy chain in a similar manner with the exception of retaining the original H3 resulted in a further increase in affinity of about 16-fold. The net increase in affinity (320-fold) resulted in a dissociation constant of 1 nM, which is comparable with the affinities obtained through hybridoma technology.

Other approaches to bypassing human immunization involve the use of semisynthetic and synthetic combinatorial libraries [Barbas et al., 1992] and the immunization of SCID mice that have been populated with human peripheral blood lymphocytes [Duchosal et al., 1992].

9.3 Monoclonal Antibody Expression Systems

Several expression systems are currently available for *in vitro* production of antibodies such as bacteria, yeast, plants, baculovirus, and mammalian cells. In each of these systems, cloned antibody genes are the starting point for production. These are obtained either by traditional cloning techniques starting with preexisting hybridomas or by the more recent repertoire cloning techniques. Each of the aforementioned systems has its own advantages and drawbacks. For example, bacterial expression systems suffer from the following limitations: They cannot be used for producing intact antibodies, nor can they glycosylate antibodies. Unglycosylated antibodies cannot perform many of the effector functions associated with normal antibody molecules. Proper folding may sometimes be a problem due to difficulty in forming disulfide bonds, and often, the expressed antibody may be toxic to the host cells. On the other hand, bacterial expression has the advantage that it is cheap, can potentially produce large amounts of the desired product, and can be scaled up easily. In addition, for therapeutic products, bacterial sources are to be preferred over mammalian sources due to the potential for the contamination of mammalian cell lines with harmful viruses.

Bacterial Expression

Early attempts to express intact antibody molecules in bacteria were fairly unsuccessful. Expression of intact light and heavy chains in the cytoplasm resulted in the accumulation of the proteins as nonfunctional inclusion bodies. *In vitro* reassembly was very inefficient [Boss et al., 1984; Cabilly et al., 1984]. These results could be explained on the basis of the fact that the *E. coli* biosynthetic environment does not support protein folding that requires specific disulfide bond formation, posttranslational modifications such as glycosylation, and polymeric polypeptide chain assembly.

There has been much more success obtained with antibody fragments. Bacterial expression of IgE "Fc-like" fragments has been reported [Kenten et al., 1984; Ishizaka et al., 1986]. These IgE fragments exhibited some of the biologic properties characteristic of intact IgE molecules. The fragments constituted 18% of the total bacterial protein content but were insoluble and associated with large inclusion bodies. Following reduction and reoxidation, greater than 80% of the chains formed dimers. The fragment binds to the IgE receptor on basophils and mast cells and, when cross-linked, elicits the expected mediator (histamine) release.

Cytoplasmic expression of a Fab fragment directed against muscle-type creatinine kinase, followed by *in vitro* folding, resulted in renaturation of about 40% of the misfolded protein, with a total active protein yield of 80 µg/ml at 10°C [Buchner and Rudolph, 1991]. Direct cytoplasmic expression of the so-called single-chain antibodies, which are novel recombinant polypeptides composed of an antibody V_L tethered to a V_H by a designed "linker" peptide that links the carboxyl terminus of the V_L to the amino terminus of the V_H or vice versa, was the next important step. Various linkers have been used to join the two variable domains. Bird et al. [1988] used linkers of varying lengths (14 to 18 amino acids) to join the two variable domains. Huston et al. [1988] used a 15 amino acid "universal" linker with the sequence $(GGGGS)_3$. The single-chain protein was found to accumulate in the cell as insoluble inclusion bodies and needed to be refolded *in vitro*. However, these proteins retained both the affinity and specificity of the native Fabs.

In 1988, two groups reported, for the first time, the expression of functional antibody fragments (Fv and Fab, respectively) in *E. coli* [Better et al., 1988; Skerra and Pluckthun, 1988]. In both cases, the authors attempted to mimic in bacteria the natural assembly and folding pathway of antibody molecules. In eukaryotic cells, the two chains are expressed separately with individual leader sequences that direct their transport to the endoplasmic reticulum (ER), where the signal sequences are removed and correct folding, disulfide formation, and assembly of the two chains occur. By expressing the two chain fragments (V_L and V_H in the case of Fv expression and the Fd and **K** chains in the case of Fab expression) separately with bacterial signal sequences, these workers were successful in directing the two precursor chains to the periplasmic space, where correct folding, assembly, and disulfide formation occur along with the removal of the signal sequences, resulting in fully functional antibody fragments. Skerra and Pluckthun [1988] report the synthesis of the Fv fragment of MOPC 603, which has an affinity constant identical to that of the intact Ab. Better et al. [1988] report the synthesis of the Fab fragment of an Ig that binds to a ganglioside antigen. While Skerra and Pluckthun obtained a yield of 0.2 mg/liter after a periplasmic wash, Better et al. found that the Fab fragment was secreted into the culture medium with a yield of 2 mg/liter. However, previous attempts to synthesize an active, full-size Ig in *E. coli* by coexpression and secretion mediated by procaryotic signal sequences resulted in poor synthesis and/or secretion of the heavy chain.

Since these early reports, several additional reports have been published (for reviews, see Pluckthun [1992] and Skerra [1993]), which have established the two aforementioned strategies (namely, direct cytoplasmic expression of antibody fragments followed by *in vitro* refolding and periplasmic expression of functional fragments) as the two standard procedures for the bacterial expression of antibody fragments. Expression of the protein in the periplasmic space has advantages: (1) the expressed protein is recovered in a fully functional form, thereby eliminating the need for *in vitro* refolding (as is required in the case of cytoplasmically expressed fragments), and (2) it greatly simplifies purification. On the other hand, direct cytoplasmic expression may, in some cases, reduce problems arising from toxicity of the expressed protein and also may increase the total yield of the protein.

Several improvements have been made in the past few years so as to simplify the expression and purification of antibody fragments in bacteria. These include the development of improved vectors with strong promoters for the high-level expression of antibody fragments, the incorporation of many different signal sequences, and the incorporation of cleavable "affinity" handles that simplify purification. Many expression vector systems are now commercially available that enable the rapid cloning, sequencing, and expression of immunoglobulin genes in bacteria within a matter of 2 to 3 weeks.

Expression in Lymphoid and Nonlymphoid Systems (Transfectoma Technology)

Expression of immunoglobulin genes by transfection into eukaryotic cells (*transfectoma technology*) such as myelomas and hybridomas is an alternative approach for producing monoclonal antibodies [Wright et al., 1992; Morrison and Oi, 1989]. Myelomas and hybridomas are known to be capable of high-level expression of endogenous heavy- and light-chain genes and can glycosylate, assemble, and secrete functional antibody molecules and therefore are the most appropriate mammalian cells for immunoglobulin gene transfection. Nonlymphoid expression in CHO and COS cells also has been examined as a potential improvement over expression in lymphoid cells. The biologic properties and effector functions, which are very important considerations for applications involving human therapy and diagnostics, are completely preserved in this mode of expression. However, transfectoma technology still involves working with eukaryotic cell lines with low antibody production rates and poor scale-up characteristics.

Transfectoma technology provides us with the ability to genetically manipulate immunoglobulin genes to produce antibody molecules with novel and/or improved properties. For example, production of "chimeric" and "reshaped" antibodies, wherein the murine variable domains or CDRs are mounted onto a human antibody framework in an attempt to reduce the problem of immunogenicity in administering murine antibodies for *in vivo* diagnostic and/or therapeutic purposes, would not be possible without the techniques of transfectoma technology. It is also possible to change the isotype of the transfectoma antibodies in order to change their biologic activity. It also has enabled the fusion of antibodies with nonimmunoglobulin proteins such as enzymes or toxins, resulting in novel antibody reagents used in industrial and medicinal applications.

The most commonly used vectors are the pSV2 vectors that have several essential features. First, they contain a plasmid origin of replication and a marker selectable for procaryotes. This makes it relatively easy to obtain large quantities of DNA and facilitates genetic manipulation. Second, they contain a marker expressible and selectable in eukaryotes. This consists of a eukaryotic transcription unit with an SV40 promoter, splice, and poly A addition site. Into this eukaryotic transcription unit is placed a dominant selectable marker derived from procaryotes (either the *neo* gene or the *gpt* gene).

In order to create the immunoglobulin molecules, the two genes encoding the heavy and light chains must be transfected and both polypeptides must be synthesized and assembled. Several methods have been used to achieve this objective. Both the heavy- and light-chain genes have been inserted into a single vector and then transfected. This approach generates large, cumbersome expression vectors, and further manipulation of the vector is difficult. A second approach is to transfect sequentially the heavy- and light-chain genes. To facilitate this, one gene is inserted into a vector with the *neo* gene, permitting selection with antibiotic G418. The other gene is placed in an expression vector containing the *gpt* gene, which confers mycophenolic acid resistance to the transfected cells. Alternatively, both genes may be introduced simultaneously into lymphoid cells using protoplast fusion.

Heavy- and light-chain genes, when transfected together, produced complete, glycosylated, assembled tetrameric antigen-binding antibody molecules with appropriate disulfide bonds. Under laboratory conditions, these transfected cells yield about 1 to 20 mg/liter of secreted antibody. A persisting problem has been the expression level of the transfected immunoglobulin gene. The expression of transfected heavy-chain genes is frequently seen to approach the level seen in myeloma cells; however, efficient expression of light-chain genes is more difficult to achieve.

Expression in Yeast

Yeast is the simplest eukaryote capable of glycosylation and secretion and has the advantages of rapid growth rate and ease of large-scale fermentation. It also retains the advantage that unlike other mammalian systems, it does not harbor potentially harmful viruses.

Initial attempts to express λ and μ immunoglobulin chains specific for the hapten NP in the yeast *Sacchromyces cerevisiae* under the control of the yeast 3-phosphoglycerate kinase (PGK) promoter resulted in the secretion into culture medium at moderate efficiency (5% to 40%), but secreted antibodies had no antigen-binding activity [Wood et al., 1985]. Subsequent attempts to coexpress heavy and light chains with the yeast invertase signal sequence under the control of the yeast phosphoglycerate kinase (PGK) promoter were more successful, presumably due to differences in the efficiency of different yeast signal sequences in directing the secretion of mammalian proteins from yeast. Culture supernatants contained significant quantities of both light and heavy chains (100 and 50 to 80 mg/liter, respectively) with about 50% to 70% of the heavy chain associated with the light chain. The yeast-derived mouse-human chimeric antibody L6 was indistinguishable from the native antibody in its antigen-binding properties [Horowitz et al., 1988; Better and Horowitz, 1989]. Furthermore, it was superior to the native antibody in mediating antibody-dependent cellular cytotoxicity (ADCC) but was incapable of eliciting complement-dependent cytolysis (CDC). Yeast-derived L6 Fab also was indistinguishable from proteolytically generated Fab as well as recombinant Fab generated from *E. coli.*

Expression in Baculovirus

The baculovirus expression system is potentially a very useful system for the production of large amounts of intact, fully functional antibodies for diagnostic and even therapeutic applications. Foreign genes expressed in insect cell cultures infected with baculovirus can constitute as much as 50% to 75% of the total cellular protein late in viral replication. Immunoglobulin gene expression is achieved by commercially available vectors that place the gene to be expressed under the control of the efficient promoter of the gene encoding the viral polyhedrin protein. The levels of expression seen in this system can be as much as 50- to 100-fold greater per cell than in procaryotes while retaining many of the advantages of an eukaryotic expression system such as glycosylation and extracellular secretion. However, scale-up is not straightforward, since viral infection eventually results in cell death.

There have been at least two reports of antibody secretion in a baculovirus system [Hasemann and Capra, 1990; Putlitz, 1990]. In both cases, the secreted antibody was correctly processed, glycosylated (albeit differently than hybridoma-derived antibody), and assembled into a normal functional heterodimer capable of both antigen binding and complement binding. The secreted antibodies were obtained at a yield of about 5 mg/liter.

Expression in Plants

The development of techniques for plant transformation has led to the expression of a number of foreign genes, including immunoglobulin genes, in transgenic plants [Hiatt and Ma, 1992]. The most commonly used plant cell transformation protocol employs the ability of plasmid Ti of *Agrobacterium tumefaciens* to mediate gene transfer into the plant genome.

The expression of a murine anti-phosphonate ester catalytic antibody in transgenic plants was accomplished by first transforming the heavy-chain and light-chain genes individually into different tobacco plants [Hiatt et al., 1989]. These were then sexually crossed to obtain progeny that expressed both chains simultaneously. It was shown that leader sequences were necessary for the proper expression and assembly of the antibody molecules. The level of antibody expression was determined to be about 3 ng/mg of total protein, and the plant-derived antibody was comparable with ascites-derived antibody with respect to binding as well as catalysis.

9.4 Genetically Engineered Antibodies and Their Fragments

The domain structure of the antibody molecule allows the reshuffling of domains and the construction of functional antibody fragments. A schematic representation of such genetically engineered antibodies and their fragments is shown in Figs. 9.2 and 9.3.

Among intact engineered constructs are chimeric, humanized, and bifunctional antibodies (see Wright et al. [1992] and Sandhu [1992] for reviews). A major issue in the long-term use of murine monoclonal antibodies for clinical applications is the immunogenicity of these molecules or the so-called HAMA response (human antimouse antibody response). Simple chimeric antibodies were constructed by linking murine variable domains to human constant domains in order to reduce immunogenicity of therapeutically administered murine monoclonals. The approach has been validated by several clinical trials that show that chimeric antibodies are much less likely to induce a HAMA response compared with their murine counterparts. In a more sophisticated approach, *CDR grafting* has been used to "humanize" murine monoclonal antibodies for human therapy by transplanting the CDRs of a murine monoclonal antibody of appropriate antigenic specificity onto a human framework. Humanized antibodies are, in some cases, even better than their chimeric counterparts in terms of the HAMA response.

Bifunctional antibodies that contain antigen-specific binding sites with two different specificities have been produced via genetic engineering as well as chemical techniques [Fanger et al., 1992]. The dual specificity of bispecific antibodies can be used to bring together the molecules or cells that mediate the desired effect. For example, a bispecific antibody that binds to target cells such as a tumor cell and to cytotoxic trigger molecules on host killer cells such as T cells has been used to redirect the normal immune system response to the tumor cells in question. Bispecific antibodies also have been used to target toxins to tumor cells.

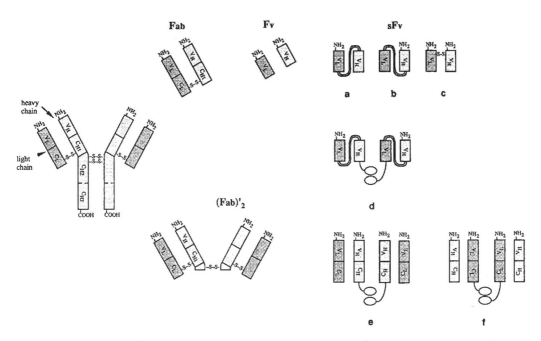

FIGURE 9.2 Schematic of an antibody molecule and its antigen binding fragments. (*a–e*) Examples of the fragments expressed in *Escherichia coli*: (*a*) single-chain Fv fragment in which the linking peptide joins the light and heavy Fv segments in the following manner: V_L (carboxyl terminus)–peptide linker–V_H (amino terminus); (*b*) single-chain Fv fragment with the following connection: V_L (amino terminus)–peptide linker–V_H (carboxyl terminus); (*c*) disulfide-linked Fv fragment; (*d*) "miniantibody" comprised of two helix-forming peptides each fused to a single-chain Fv; (*e*) Fab fragments linked by helix-forming peptide fused to heavy chain; (*f*) Fab fragments linked by helix-forming peptide to light chain.

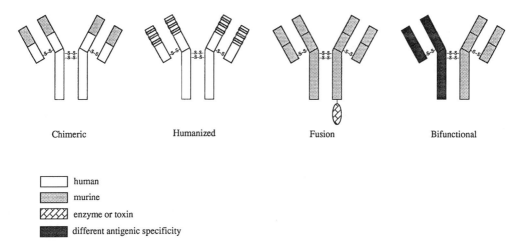

FIGURE 9.3 Schematic of genetically engineered intact monoclonal antibodies prepared with antibody-engineering techniques.

The list of genetically engineered antibody fragments is a long and growing one [Pluckthun, 1992]. Of these, fragments such as the Fab, F(ab)$'_2$, Fv, and the Fc are not new and were initially produced by proteolytic digestion. The Fab fragment (fragment antigen-binding) consists of the entire light chain and the two N-terminal domains of the heavy chain (V_H and C_{H1}, the so-called Fd chain) and was first generated by digestion with papain. The other fragment that is generated on papain digestion is the Fc fragment (fragment crystallizable), which is a dimeric unit composed of two C_{H2} and C_{H3} domains. The $F(ab)'_2$ fragment consists of two Fab arms held together by disulfide bonds in the hinge region and was first generated by pepsin digestion. Finally, the Fv fragment, which consists of just the two N-terminal variable domains, also was first generated by proteolytic digestion and is now more commonly generated via antigen-engineering techniques.

The other fragments listed in Fig. 9.2 are of genetic origin. These include the sFv (single-chain antigen), the V_H domain (single-domain antigen), multivalent sFvs (miniantibodies), and multivalent Fabs. The multivalent constructs can either be monospecific or bispecific. The single-chain antigen (SCA, sFv) consists of the two variable domains linked together by a long flexible polypeptide linker [Bird et al., 1988; Huston et al., 1988]. The sFv is an attempt to stabilize the Fv fragment, which is known to dissociate at low concentrations into its individual domains due to the low-affinity constant for the V_H-V_L interaction. Two different constructs have been made: the V_L-V_H construct where the linker extends from the C terminus of the V_L domain to the N terminus of the V_H domain, and the V_H-V_L construct, where the linker runs from the C terminus of the V_H domain to the N terminus of the V_L. The linker is usually about 15 amino acids long (the length required to span the distance between the two domains) and has no particular sequence requirements other than to minimize potential interferences in the folding of the individual domains. The so-called universal linker used by many workers in the area is $(GGGGS)_3$. Other strategies to stabilize the Fv fragment include chemical cross-linking of the two domains and disulfide-linked domains [Glockshuber et al., 1990]. Chemical cross-linking via glutaraldehyde has been demonstrated to be effective in stabilizing the Fv fragment in one instance; here, the cross-linking was carried out in the presence of the hapten (phosphorylcholine) to avoid modification of the binding site, an approach that may not be feasible with protein antigens. In the disulfide-linked sFv, Cys residues are introduced at suitable locations in the framework region of the Fv so as to form a natural interdomain disulfide bond. This strategy was shown to be much more effective in stabilizing the Fv fragment against thermal denaturation than either the single-chain antigen approach or chemical cross-linking for the one case where all three approaches were tested [Glockshuber et al., 1990].

The single-domain antigen consists of just the V_H domain and has been shown by some to possess antigen-binding function on its own in the absence of the V_L domain [Ward et al., 1989]. There is some skepticism regarding this approach due to the rather high potential for nonspecific binding (the removal of the V_L domain exposes a very hydrophobic surface), poor solubility, and somewhat compromised selectivity. For example, while the Fv fragment retains its ability to distinguish between related antigenic species, the single-domain antigen does not.

Miniantibodies consist of sFv fragments held together by so-called dimerization handles. sFv fragments are fused via a flexible hinge region to several kinds of amphipathic helices, which then acts as dimerization devices [Pack and Pluckthun, 1992]. Alternately, the two sFvs can be fused via a long polypeptide linker, similar to the one linking the individual domains of each Fv but longer to maintain the relative orientation of the two binding sites. Multivalent Fabs use a somewhat similar approach with the dimerization handles comprising of zippers from the transcription factors *jun* and *fos* [Kostelny et al., 1992].

In addition to the constructs described above, a whole new set of genetically engineered fusion proteins with antibodies has been described [Wright et al., 1992]. Antibody fusion proteins are made by replacing a part of the antibody molecule with a fusion partner such as an enzyme or a toxin that confers a novel function to the antibody. In some cases, such as immunotoxins, the variable regions of the antibody are retained in order to retain antigen binding and specificity, while the constant domains are deleted and replaced by a toxin such as ricin or *Pseudomonas* exotoxin. Alternately, the constant regions are retained (thereby retaining the effector functions) and the variable regions replaced with other targeting proteins (such as CD4 for AIDS therapy and IL-2 for cancer therapy).

9.5 Applications of Monoclonal Antibodies and Fragments

The majority of applications for which monoclonal antibodies have been used can be divided into three general categories: (1) purification, (2) diagnostic functions (whether for detecting cancer, analyzing for toxins in food, or monitoring substance abuse by athletes), and (3) therapeutic functions. From the time that monoclonal antibody technology was introduced almost 20 years ago, application methodologies using whole antibody have gradually been transformed into methodologies using antibody fragments such as the Fab_2, Fab′, and Fv fragments and even synthetic peptides of a CDR region. Antibody conjugates have come to include bound drugs, toxins, radioisotopes, lymphokines, and enzymes [Pietersz and McKenzie, 1992] and are largely used in the treatment of cancer. Tables 9.1 and 9.2 list some typical examples of monoclonal antibodies and fragments used for diagnostic and therapeutic applications.

TABLE 9.1 Uses of Whole MAb Derived from Hybridoma, Other Cell Fusions, and Genetically Engineered Cell Lines

MAb	Antigen	Ab Source	Actual or Potential Uses
C23	Cytomegalovirus glycoprotein of 130 and 55 kDa	Humab — fusion of human lymphocyte and mouse myeloma (p3 × 63Ag8ul)	Prophylactic agent for viral infection
EV2-7	Cytomegalovirus protein of 82 kDa	Trioma: human × (human × mouse heteromyeloma)	Prophylactic agent for CMV infection
OKT3	CD3 antigen	Murine MAb	Eradication of T lymphocytes involved in graft rejection
Campath 1H	Lymphocyte antigen	"Humanized" chimeric rat/human MAb; murine CDRs	Prevention of bone marrow and organ rejection
R24	GD2 ganglioside TADA tumor-associated differentiation antigen	Murine MAb	Treatment of melanoma
L72, L55	GD3 ganglioside TADA	Human MAb	Treatment of melanoma
	Digoxin	Murine MAb	Immunodiagnostic for cardiac glycoside digoxin
6H4	*Salmonella* flagella	Murine MAb	Detection of *Salmonella* bacteria in food

TABLE 9.2 Immunoconjugates Having Potential for Cancer Therapy

MAb	Conjugate	Antibody Source	Use
A7	Neocarzinostatin	Murine monoclonal (from fusion with murine myeloma P3, X63.Ag8.653)	Eradication of colon cancer
30.6, I-1	*N*-acetylmelphalan	Murine monoclonal	Eradication of colon cancer
	Adriamycin, Mitomycin C	Murine monoclonal	Various cancers
RFB4	RFB4(Fab′)-Ricin A (αCD22 (Fab′)-Ricin A)	Murine Fab′	B-cell lymphoma
	Anti-CD19-blocked ricin	Murine monoclonal	B-cell lymphoma
	Xomazyme-Mel	Murine monoclonal	Metastatic melanoma
Anti-CEA (carcinoembryonic antigen)	Carboxypeptidase G2	Murine MAb F(ab′)$_2$ fragment chemically bound to enzyme	Colon cancer
Recombinant anti-CEA MAb BW431	(DNA coded) human β-glucuronidase	Transfectoma	
B72.3	^{131}I	Murine monoclonal	Ovarian cancer
B72.3	^{131}I	Mouse-human chimeric Ab	Colon cancer
B72.3		Mouse-human chimeric Fab′ fragment	Colon cancer
B72.3 (Oncoscint)	Chelated ^{111}In	Murine MAb	Diagnostic imaging agent for colorectal and ovarian cancers

Thousands of murine monoclonal antibodies have been made to human carcinomas since the introduction of antibody technology, but very few, if any, of these monoclonal antibodies are entirely specific for malignant cells. In the vast majority of cases, these monoclonal antibodies define tumor-associated differentiation antigens (TADAs), which are either proteins, mucins, proteoglycans, or glycolipids (gangliosides). Examples of TADA proteins are carcinoembryonic antigen (CEA) and α-fetoprotein (AFP), both well-known diagnostic markers. An anti-CEA antibody has been conjugated with the enzyme carboxypeptidase, which, in turn, activates a prodrug at the site of the tumor [Bagshawe et al., 1992]. This strategy overcomes the inability of monoclonal antibodies conjugated with drugs to deliver a therapeutic dose. Examples of TADA gangliosides are those referred to as GD2 and GD3, for which the respective unmodified monoclonal antibodies have shown to be effective therapeutics, particularly when intralesionally administered [Irie et a., 1989]. An example of a TADA mucin is the antigens found in colorectal and ovarian carcinoma that react with the antibody 72.3. Chimeric monoclonal antibodies as well as fragments of antibody 72.3 have been constructed and tested [Khazaeli et al., 1991; King et al., 1992].

As mentioned above, monoclonal antibodies defining different TADAs have been used passively (unmodified) and as carriers of, for example, radioisotopes and enzymes. From the results brought forth to date, the passive mode of antibody therapy has produced relatively few remissions in patients, and in those cases where it has shown effects, it is likely that the ability of the antibody to mediate ADCC (antibody-dependent cellular cytotoxicity) and CDC (complement-dependent cytotoxicity) has contributed to the remission. For the case of modified monoclonal antibodies (e.g., toxin and radioisotope conjugates), success of treatment is varied depending on the type of neoplasm. Antibody conjugate treatment of leukemia and lymphoma results in a relatively greater remission rate than that found in treatments of malignancies having solid tumors (i.e., carcinoma of the ovary, colon, and lung).

The rare case of complete remission for solid tumors is probably due to the inaccessibility of antibody to that tumor. Several barriers impeding access of antibody to cancer cells have been pointed out [Jain, 1988]. A few of these barriers include (1) the high interstitial fluid pressure in tumor nodules, (2) heterogeneous or poor vascularization of tumors, and (3) the long distances extravasated monoclonal antibodies must travel in the interstitial mesh of proteoglycans in the tumor. There also exists the possibility that tumor antigen shed from the surface is limiting antibody buildup. In the case of bound toxins or drugs, there is the added concern that organs such as the kidney and liver are quickly processing and eliminating the antibody conjugates. In this respect, immunoconjugates based on antibody fragments (such as the sFv) can be very advantageous. For example, it has been shown that the sFv exhibits rapid diffusion into the extravascular space, increased volume of distribution, enhance tumor penetration, and improved renal elimination. An assessment of solid tumor therapy with modified antibody has led Riethmüller et al. [1993] to recommend that current cancer therapy be directed toward minimal residual disease, the condition in which micrometastatic cells exist after curatively resecting solid tumors.

With regard to purification, the research literature is replete with examples of immunoaffinity purification of enzymes, receptors, peptides, and small organic molecules. In contrast, commercial applications of immunoaffinity chromatography, even on industrially or clinically relevant molecules, are far less widespread (Table 9.3). Despite its potential utility, immunoadsorption is an expensive process. A significant portion of the high cost is the adsorbent itself, which is related to the cost of materials, preparation, and most important, the antibody. In addition, the binding capacity of the immunoadsorbent declines with repeated use, and a systemic study has shown that significant deactivation can typically take place over 40 to 100 cycles [Antonsen et al., 1991]. A number of factors can contribute to this degradation, including loss of antibody, structural change of the support matrix, nonspecific absorption of contaminating proteins, incomplete antigen elution, and loss of antibody function. In most cases, this degradation is associated with repeated exposure to harsh elution conditions. Noteworthy commercial applications of immunoaffinity chromatography on useful molecules include the separation of factor VIII used to treat hemophilia A and factor IX, another coagulation factor in the blood-clotting cascade [Tharakan et al., 1992]. The immunoaffinity purification step for factor VIII was one of several additional steps added to the conventional preparation methodology in which plasma cryoprecipitates were heat-treated. The new method contains a virus-inactivation procedure that precedes the immunoaffinity

TABLE 9.3   Clinically or Industrially Relevant Proteins Purified by Immunoaffinity Chromatography

Protein	Use (Actual and Potential)
Factor VIII	Treatment of hemophilia A
Factor IX	Blood coagulant
α-Galactosidase	Improve the food stabilizing properties of guar gum
Alkaline phosphatase	Purify enzyme (a particular glycoform) used as a tumor marker for diagnostic tests
Interferon (recombinant)	Immunotherapeutic
Interleukin 2	Immunotherapeutic

column, followed by an additional chromatographic step (ion exchange). The latter step serves to eliminate the eluting solvent and further reduce virus-inactivating compounds. The often-mentioned concern of antibody leakage from the column matrix did not appear to be a problem. Furthermore, with the relatively mild elution conditions used (40% ethylene glycol), one would expect little change in the antibody-binding capacity over many elution cycles.

Typical immunoaffinity matrices contain whole antibody as opposed to antibody fragments. Fragmentation of antibody by enzymatic means contributes additional steps to immunoadsorbent preparation and adds to the overall cost of the separation and is thus avoided. However, fragmentation can lead to a more efficient separation by enabling the orientation of antibody-binding sites on the surface of the immunomatrix [Prisyazhnoy et al., 1988; Yarmush et al., 1992]. Intact antibodies are bound in a random fashion, resulting in a loss of binding capacity upon immobilization. Recombinant antibody fragments could prove to be more useful for immunoaffinity applications due to the potential for production of large quantities of the protein at low cost and improved immobilization characteristics and stability [Spitznagel and Clark, 1993]. In what one could consider the ultimate fragment of a antibody, some investigators have utilized a peptide based on the CDR region of one of the chains (termed *minimal recognition units*) to isolate the antigen. Welling et al. [1990] have synthesized and tested a 13-residue synthetic peptide having a sequence similar to one hypervariable region of an antilysozyme antibody.

Important diagnostic uses of antibody include the monitoring in clinical laboratories of the cardiac glycoside digoxin and the detection of the *Salmonella* bacteria in foods (Table 9.1). These two examples highlight the fact that despite the exquisite specificity offered by monoclonal antibodies, detection is not failure-proof. Within digoxin immunoassays there are two possible interfering groups: endogenous digoxin-like substances and digoxin metabolites; moreover, several monoclonal antibodies many be necessary to avoid under- or overestimating digoxin concentrations. In the case of bacteria detection in food, at least two antibodies (MOPC 467 myeloma protein and 6H4 antibody) are needed to detect all strains of *Salmonella*.

9.6   Summary

The domain structure of antibodies, both at the protein and genetic levels, facilitates the manipulation of antibody properties via genetic engineering (antibody engineering). Antibody engineering has shown tremendous potential for basic studies and industrial and medical applications. It has been used to explore fundamental questions about the effect of structure on antigen binding and on the biologic effector functions of the antibody molecules. A knowledge of the rules by which the particular sequences of amino acids involved in the binding surface are chosen in response to a particular antigenic determinant would enable the production of antibodies with altered affinities and specificities. Understanding the structures and mechanisms involved in the effector function of antibodies is already starting to result in the production of antibodies with novel biologic effector functions for use as diagnostic and therapeutic reagents. In addition, the production of antibodies via immunoglobulin gene expression has enabled the engineering of novel hybrid, chimeric, and mosaic genes using recombinant DNA techniques and the transfection and expression of these genetically engineered genes in a number of different systems such as bacteria and yeast, plant cells, myeloma or hybridoma cells, and nonlymphoid mammalian cells.

References

Antonsen KP, Colton CK, Yarmush ML. 1991. Elution conditions and degradation mechanisms in long-term immunoadsorbent use. *Biotechnol Prog* 7:159.

Bagshawe KD, Sharma SK, Springer CJ, et al. 1992. Antibody directed enzyme prodrug therapy (ADEPT). *Antibody Immunoconj Radiopharm* 54:133.

Barbas CF, Kang AS, Lerner RA, Benkovic SJ. 1991. Assembly of combinatorial antibody libraries on phase surfaces: The gene III site. *Proc Natl Acad Sci USA* 88:7978.

Barbas CF, Bain JD, Hoekstra DM, Lerner RA. 1992. Semisynthetic combinatorial antibody libraries: A chemical solution to the diversity problem. *Proc Natl Acad Sci USA* 89:4457.

Bender E, Woof JM, Atkin JD, et al. 1993. Recombinant human antibodies: Linkage of an Fab fragment from a combinatorial library to an Fc fragment for expression in mammalian cell culture. *Hum Antibod Hybridomas* 4:74.

Better M, Chang CP, Robinson RR, Horwitz AH. 1988. *Escherichia coli* secretion of an active chimeric antibody fragment. *Science* 240:1041.

Better M, Horowitz AH. 1989. Expression of engineered antibodies and antibody fragments in microorganisms. *Methods Enzymol* 178:476.

Bird RE, Hardman KD, Jacobson JW, et al. 1988. Single-chain antibody-binding proteins. *Science* 242:423.

Boss MA, Kenten JH, Wood CR, Emtage JS. 1984. Assembly of functional antibodies from immunoglobulin heavy and light chains synthesized in *E. coli. Nucleic Acids Res* 12:3791.

Buchner J, Rudolph R. 1991. Renaturation, purification, and characterization of recombinant Fab fragments produced in *Escherichia coli. Biotechnology* 9:157.

Burton DR. 1991. Human and mouse monoclonal antibodies by repertoire cloning. *Trends Biotechnol* 9:169.

Burton DR, Barbas CF, Persson MAA, et al. 1991. A large array of human monoclonal antibodies to type 1 human immunodeficiency virus from combinatorial libraries of asymptomatic seropositive individuals. *Proc Natl Acad Sci USA* 88:10134.

Cabilly S, Riggs AD, Pande H, et al. 1984. Generation of antibody activity from immunoglobulin polypeptide chains produced in *Escherichia coli. Proc Natl Acad Sci USA* 81:3273.

Caton AJ, Koprowski H. 1990. Influenza virus hemagglutinin-specific antibodies isolated from a combinatorial expression library are closely related to the immune response of the donor. *Proc Natl Acad Sci USA* 87:6450.

Clackson T, Hoogenboom HR, Griffiths AD, Winter G. 1991. Making antibody fragments using phage display libraries. *Nature* 352:624.

Davies DR, Padlan EA, Sheriff S. 1990. Antigen-antibody complexes. *Annu Rev Biochem* 59:439.

Duchosal MA, Eming SA, Fischer P, et al. 1992. Immunization of hu-PBL-SCID mice and the rescue of human monoclonal Fab fragments through combinatorial libraries. *Nature* 355:258.

Fanger MW, Morganelli PM, Guyre PM. 1992. Bispecific antibodies. *Crit Rev Immunol* 12:101.

Glockshuber R, Malia M, Pfitzinger I, Pluckthun A. 1990. A comparison of strategies to stabilize immunoglobulin fragments. *Biochemistry* 29:1362.

Gram H, Lore-Anne M, Barbas CF, et al. 1992. In vitro selection and affinity maturation of antibodies from a naive combinatorial immunoglobulin library. *Proc Natl Acad Sci USA* 89:3576.

Hassemann CA, Capra JD. 1990. High-level production of a functional immunoglobulin heterodimer in a baculovirus expression system. *Proc Natl Acad Sci USA* 87:3942.

Hiatt A, Cafferkey R, Bowdish K. 1989. Production of antibodies in transgenic plants. *Nature* 342:76.

Hiatt A, Ma JK-C. 1992. Monoclonal antibody engineering in plants. *FEBS Lett* 307:71.

Hoogenboom HR, Griffiths AD, Johnson KS, et al. 1991. Multi-subunit proteins on the surface of filamentous phage: Methodologies for displaying antibody (Fab) heavy and light chains. *Nucleic Acids Res* 19:4133.

Hoogenboom HR, Marks JD, Griffiths AD, Winter G. 1992. Building antibodies from their genes. *Immunol Rev* 130:41.

Horowitz AH, Chang PC, Better M, et al. 1988. Secretion of functional antibody and Fab fragment from yeast cells. *Proc Natl Acad Sci USA* 85:8678.

Huse WD, Sastry L, Iverson SA, et al. 1989. Generation of a large combinatorial library of the immuno-globulin repertoire in phage lambda. *Science* 246:1275.

Huston JS, Levinson D, Mudgett-Hunter M, et al. 1988. Protein engineering of antibody binding sites: Recovery of specific activity in an anti-digoxin single-chain Fv analogue produced in *Escherichia coli*. *Proc Natl Acad Sci USA* 85:5879.

Ishizaka T, Helm B, Hakimi J, et al. 1986. Biological properties of a recombinant human immunoglobulin ε-chain fragment. *Proc Natl Acad Sci USA* 83:8323.

Jain RK. 1988. Determinants of tumor blood flow: A review. *Cancer Res* 48:2641.

Kang AS, Barbas CF, Janda KD, et al. 1991. Linkage of recognition and replication functions by assembling combinatorial antibody Fab libraries along phage surfaces. *Proc Natl Acad Sci USA* 88:4363.

Keneten J, Helm B, Ishizaka T, et al. 1984. Properties of a human immunoglobulin ε-chain fragment synthesized in *Escherichia coli*. *Proc Natl Acad Sci USA* 81:2955.

Khazaeli MB, Saleh MN, Liu TP, Meredith RF. 1991. Pharmacokinetics and immune response of 131I-chi-meric mouse/human B72.3 (human γ4) monoclonal antibody in humans. *Cancer Res* 51:5461.

King DJ, Adair JR, Angal S, et al. 1992. Expression, purification, and characterization of a mouse-human chimeric antibody and chimeric Fab′ fragment. *Biochem J* 281:317.

Kostelny SA, Cole MS, Tso JY. 1992. Formation of bispecific antibody by the use of leucine zippers. *J Immunol* 148:1547.

Köhler G, Milstein C. 1975. Continuous cultures of fused cells secreting antibody of predefined specificity. *Nature* 256:495.

Landsteiner K. 1945. *The Specificity of Serological Reactions*. Cambridge, Mass, Harvard University Press.

Marks JD, Hoogenboom HR, Bonnert TP, et al. 1991. Bypassing immunization: Human antibodies from V-gene libraries displayed on phase. *J Mol Biol* 222:581.

Marks JD, Griffiths AD, Malmqvist M, et al. 1992. Bypassing immunization: Building high affinity human antibodies by chain shuffling. *Biotechnology* 10:779.

McCafferty J, Griffiths AD, Winter G, Chriswell DJ. 1990. Phage antibodies: Filamentous phage displaying antibody variable domains. *Nature* 348:552.

Morrison SL, Oi VT. 1989. Genetically engineered antibody molecules. *Adv Immunol* 41:65.

Mullinax RL, Gross EA, Amberg JR, et al. 1990. Identification of human antibody fragment clones specific for tetanus toxoid in a bacteriophage λ immunoexpression library. *Proc Natl Acad Sci USA* 87:8095.

Neil GA, Urnovitz HB. 1988. Recent improvements in the production of antibody-secreting hybridoma cells. *Trends Biotechnol* 6:209.

Orlandi R, Gussow DH, Jones PT, Winter G. 1989. Cloning immunoglobulin variable domains for expression by the polymerase chain reaction. *Proc Natl Acad Sci USA* 86:3833.

Pack P, Pluckthun P. 1992. Miniantibodies: Use of amphipathic helices to produce functional flexibility linked dimeric Fv fragments with high avidity in *Escherichia coli*. *Biochemistry* 31:1579.

Perrson MAA, Caothien RH, Burton DR. 1991. Generation of diverse high-affinity human monoclonal antibodies by repertoire cloning. *Proc Natl Acad Sci USA* 88:2432.

Pietersz GA, McKenzie IFC. 1992. Antibody conjugates for the treatment of cancer. *Immunol Rev* 129:57.

Plückthun A. 1992. Mono- and bivalent antibody fragments produced in Escherichia coli: Engineering, folding and antigen-binding. *Immunol Rev* 130:150.

Prisyazhnoy VS, Fusek M, Alakhov YB. 1988. Synthesis of high-capacity immunoaffinity sorbents with oriented immobilized immunoglobulins or their Fab′ fragments for isolation of proteins. *J Chromatogr* 424:243.

Putlitz JZ, Kubasek WL, Duchene M, et al. 1990. Antibody production in baculovirus-infected insect cells. *Biotechnology* 8:651.

Riethmüller G, Schneider-Gädicke E, Johnson JP. 1993. Monoclonal antibodies in cancer therapy. *Curr Opin Immunol* 5:732.

Sandhu JS. 1992. Protein engineering of antibodies. *Crit Rev Biotechnol* 12:437.

Sastry L, Alting-Mees M, Huse WD, et al. 1989. Cloning of the immunoglobulin repertoire in *Escherichia coli* for generation of monoclonal catalytic antibodies: Construction of a heavy chain variable region-specific cDNA library. *Proc Natl Acad Sci USA* 86:5728.

Skerra A, Plückthun A. 1988. Assembly of a functional immunoglobulin Fv fragment in *Escherichia coli*. *Science* 240:1038.

Skerra A. 1993. Bacterial expression of immunoglobulin fragments. *Curr Opin Biotechnol* 5:255.

Spitznagel TM, Clark DS. 1992. Surface density and orientation effects on immobilized antibodies and antibody fragments. *Biotechnology* 11:825.

Tizard IR. 1992. The genetic basis of antigen recognition. In *Immunology: An Introduction*, 3d ed. Orlando, Fla. Saunders Coolege Publishing.

Ward ES, Gussow D, Griffiths AD, et al. 1989. Binding activities of a repertoire of single immunoglobulin variable domains secreted from *Escherichia coli. Nature* 341:544.

Wood CR, Boss MA, Kenten JH, et al. 1985. The synthesis and in vivo assembly of functional antibodies in yeast. *Nature* 314:446.

Welling GW, Guerts T, Van Gorkum J, et al. 1990. Synthetic antibody fragment as ligand in immuno-affinity chromatography. *J. Chromatogr* 512:337.

Wright A, Shin S-U, Morrison SL. 1992. Genetically engineered antibodies: progress and prospects. *Crit Rev Immunol* 12:125.

Yarmush ML, Lu X, Yarmush D. 1992. Coupling of antibody-binding fragments to solid-phase supports: Site-directed binding of F(ab′)2 fragments. *J Biochem Biophys Methods* 25:285.

Zebedee SL, Barbas CF, Yao-Ling H, et al. 1992. Human combinatorial libraries to hepatitis B surface antigen. *Proc Natl Acad Sci USA* 89:3175.

10

Antisense Technology

S. Patrick Walton
Center for Engineering in Medicine,
Massachusetts General Hospital,
Harvard Medical School, and
Shriners Burns Hospital, Boston

Charles M. Roth
Center for Engineering in Medicine,
Massachusetts General Hospital,
Harvard Medical School, and
Shriners Burns Hospital, Boston

Martin L. Yarmush
Center for Engineering in Medicine,
Massachusetts General Hospital,
Harvard Medical School, and
Shriners Burns Hospital, Boston

Antisense molecules can selectively inhibit the expression of one gene among the 100,000 present in a typical human cell. This inhibition is likely based on simple Watson-Crick base-pairing interactions between nucleic acids and makes possible, in principle, the rational design of therapeutic drugs that can specifically inhibit any gene with a known sequence. The intervention into disease states at the level of gene expression may potentially make drugs based on antisense techniques significantly more efficient and specific than other standard therapies. Indeed, antisense technology is already an indispensable research tool and may one day be an integral part of future antiviral and anticancer therapies.

Zamecnik and Stephenson were the first to use antisense DNA to modulate gene expression. They constructed antisense oligonucleotides complementary to the 3′ and 5′ ends of Rous sarcoma virus (RSV) 35S RNA and added them directly to a culture of chick embryo fibroblasts infected with RSV [1]. Remarkably, viral replication was inhibited, indicating that the cells had somehow internalized the antisense DNA and that the DNA was somehow interrupting the viral life cycle.

Natural antisense inhibition was first observed in bacteria as a means of regulating the replication of plasmid DNA [2, 3]. RNA primers required for the initiation of replication were bound by (i.e., formed duplexes with) antisense RNA. The concentration of these RNA primers, and therefore the initiation of replication, was controlled by the formation of these duplexes. Shortly after this discovery, investigators developed antisense RNA constructs to control gene expression in mammalian cells. Antisense RNA, encoded on expression plasmids that were transfected into mouse cells, successfully blocked expression of target genes [4]. These early successes launched what is now a significant effort to expand the use of antisense molecules for research and therapeutic purposes.

10.1 Background

Antisense oligonucleotides are DNA or RNA molecules whose sequences are complementary to RNA transcribed from the target gene. These molecules block the expression of a gene by interacting with

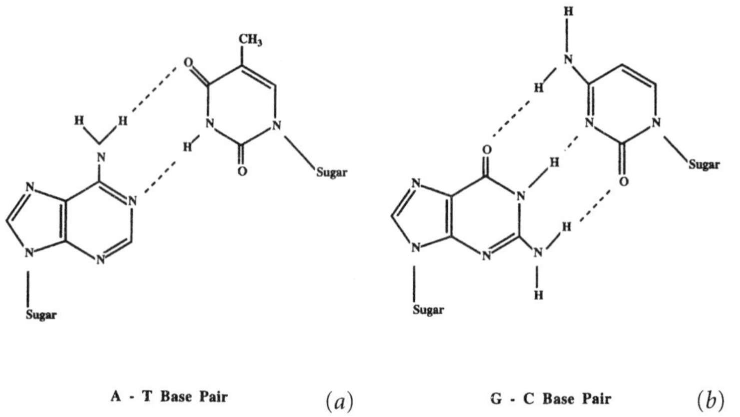

A - T Base Pair   (*a*)   G - C Base Pair   (*b*)

FIGURE 10.1 Watson-Crick base pairing interactions between adenosine and thymidine (A-T) and between guanosine and cytidine (G-C). The sugar is $2'$-deoxyribose.

its RNA transcript. Antisense molecules are typically short, single-stranded oligonucleotides with a sequence complementary to a sequence within the target RNA transcript. The oligonucleotides bind to this sequence via Watson-Crick base pairing (adenosine binds to thymidine (DNA) or uracil (RNA), and guanosine binds to cytidine), form a DNA:RNA duplex, and block translation of the RNA (see Fig. 10.1).

Nucleic acids have been used in a variety of strategies to modulate gene expression. For example, antisense RNA, encoded by a plasmid transfected into the target cells, has been used to inhibit gene expression. Cells are transfected with a plasmid encoding the antisense RNA, the plasmid is transcribed, and the resulting antisense RNA transcript inhibits gene expression. Unfortunately, difficulties in controlling the expression of the transfected plasmid hinder the effective use of antisense RNA for therapeutic purposes. More recently, antisense RNA work has focused on the use of hammerhead ribozymes. These synthetic RNA oligonucleotides have the ability to cleave other RNA strands and they appear more therapeutically viable. Other methods for using nucleic acids to block gene expression are being explored, such as:

1. nucleic acids that competitively bind to proteins via their 3-dimensional structure (aptamers),
2. nucleic acids designed to prevent transcription by forming a DNA triplex with the target gene (antigene), and
3. nucleic acids that mimic binding sites for transcription and translation complexes (decoys).

Although promising, these approaches are preliminary and are beyond the scope of this chapter. The interested reader is directed to reviews concerning these topics [5-7]. Only antisense DNA molecules, which mimic the antisense strand of the target gene and thus hybridize to RNA transcribed from the gene, will be discussed in detail here. For a recent review on antisense RNA, see Rossi [8].

Many technical issues limit the therapeutic usefulness of current antisense oligonucleotides; for instance, they are highly susceptible to degradation by nucleases. Also, our understanding of the mechanism of antisense inhibition must improve before the widespread development of therapeutically useful antisense molecules is possible. Advances in oligonucleotide chemistry have begun to resolve the issue of their intracellular stability; however, the reaction of natural systems to these unnatural species has not been fully explored. Many oligonucleotides still lack adequate specificity for *in vivo* use, and their delivery to cells is often non-specific and inefficient. The impact of these problems and approaches to solving them will be discussed. In addition, potential applications of antisense techniques to antiviral and anticancer therapies and their evaluation in animal models and clinical trials will also be highlighted.

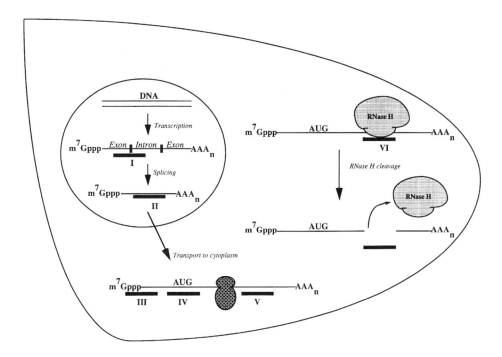

FIGURE 10.2 Several possible sequence specific sites of antisense inhibition. Antisense oligodeoxynucleotides are represented by black bars. Antisense oligodeoxynucleotides can interfere with (I) splicing, (II) transport of the nascent mRNA from the nucleus to the cytoplasm, (III) binding of initiation factors, (IV) assembly of ribosomal subunits at the start codon or (V) elongation of translation. Inhibition of capping and polyadenylation is also possible. (VI) Antisense oligodeoxynucleotides that activate RNase H (e.g., oligonucleotides with phosphodiester and phosphorothioate backbones) can also inhibit gene expression by binding to their target mRNA, catalyzing the RNase H cleavage of the mRNA into segments that are rapidly degraded by exonucleases.

10.2 Mechanisms of Inhibition

The inhibition of gene expression by antisense molecules is believed to occur by a combination of two mechanisms: ribonuclease H (RNase H) degradation of the RNA and steric hindrance of the processing of the RNA [9]. Some oligonucleotides, after hybridizing to the target RNA, activate RNase H, which specifically recognizes an RNA:DNA duplex and cleaves the RNA strand. The antisense oligonucleotide is not degraded and is free to bind to and catalyze the destruction of other target RNA transcripts [10]. Oligonucleotides that activate RNase H, therefore, are potentially capable of destroying many RNA transcripts in their lifetime, which suggests that lower concentrations of these oligonucleotides may be sufficient to significantly inhibit gene expression [9].

However, not all oligonucleotides activate RNase H enzymes. Some oligonucleotides inhibit gene expression by interfering with the RNA's normal life cycle. Newly transcribed RNA must be spliced, polyadenylated, and transported to the cytoplasm before being translated into protein. In theory, virtually any step in an RNA's life cycle can be blocked by an oligonucleotide (see Fig. 10.2). Oligonucleotides that do not activate RNase H can block RNA maturation by binding to sites on the RNA important for post-transcriptional processing or for ribosomal assembly and translation initiation [11]. Oligonucleotides designed to block the elongation of translation by binding to coding sequences downstream from the initiation codon region have rarely succeeded, presumably because ribosomes destabilize and read through DNA:RNA duplexes [11]. Inhibition by both RNase H degradation and steric hindrance was demonstrated in a study in which the expression of an intracellular adhesion molecule (ICAM-1) was blocked using two different oligonucleotides [12].

The mechanism of inhibition influences the specificity of antisense molecules. Oligonucleotides that activate RNase H can inactivate slightly mismatched RNA transcripts by transiently binding to them and catalyzing their cleavage and destruction by RNase H [13]. This nonspecific inactivation of RNA could lead to unwanted side effects. In contrast, oligonucleotides that do not activate RNase H, and therefore can only inhibit gene expression by steric hindrance, are less likely to inactivate slightly mismatched RNA transcripts. These oligonucleotides, like the ones that activate RNase H, transiently bind to RNA transcripts. However, they do not catalyze the enzymatic cleavage of the RNA by RNase H, and, since transient binding is unlikely to significantly impede translation of the transcript, they are less likely to reduce the expression of slightly mismatched nontargeted RNA transcripts. [14].

Choosing an effective target site in an mRNA remains largely an empirical process. Generally, several candidate oligonucleotides are tested for their ability to inhibit gene expression, and the most effective one is used for future studies [10]. Techniques such as gel-shift or oligonucleotide array assays are currently being used to rapidly screen multiple candidate oligonucleotides *in vitro* [15, 16]. Computational predictions have also been applied to assist in the identification of susceptible sites on the target RNA [17, 18]. The effectiveness of an antisense oligonucleotide has been shown to depend greatly on the region of the RNA to which it is complementary [19, 20].

RNA molecules are typically folded into complex 3-dimensional structures that are formed by self-base pairing and by protein-RNA interactions. If an oligonucleotide is complementary to a sequence embedded in the 3-dimensional structure, the oligonucleotide may not be able to bind to its target site, and, therefore, gene expression will not be inhibited. For oligonucleotides that block gene expression only by steric hindrance and not by RNase H degradation, the number of potential target sites is even more limited. For these oligonucleotides, the site must not only be accessible for hybridization, but it must be a site for which steric hindrance alone is sufficient to interfere with gene expression. Binding sites that are often effective include the 5′ cap region and the initiation (AUG) codon region [15, 21-23]. However, in other cases, effective inhibitors proved to be oligonucleotides targeted to segments of the RNA distant from these regions, e.g., in the 3′-UTR of the mRNA [20, 23].

10.3 Chemically Modified Oligonucleotides

Natural oligonucleotides with a phosphodiester backbone (PO oligonucleotides) are very susceptible to nuclease degradation, having half-lives in the presence of purified exonucleases as short as fifteen minutes [11, 24, 25]. Almost no full sequence PO oligonucleotide remained following a 1-h incubation in human serum or cell extracts [26]. The need for oligonucleotides that are stable *in vivo* prompted the development of chemical derivatives of the phosphodiester backbone. The first derivatives were methylphosphonate and phosphorothioate backbone oligonucleotides. Methylphosphonate oligonucleotides have a methyl group in place of one of the nonbridge backbone oxygens. They are highly lipophilic with a neutral backbone. Recently, however, little attention has been afforded to methylphosphonates as they appear less therapeutically viable due to their poor cellular uptake and the inability of their duplexes to be recognized by RNase H [10, 11].

Phosphorothioate oligonucleotides, on the other hand, remain prominent. Phosphorothioate (PS) DNA derivatives have a sulfur atom substituted for a nonbridge backbone oxygen atom. PS oligonucleotides are significantly more stable than PO oligonucleotides with a half-life of >5 h in the presence of purified nucleases [25]. Like PO oligonucleotides, PS oligonucleotides are believed to be internalized by receptor mediated endocytosis [27]. Their affinity for complementary RNA is not as high as that observed with PO oligonucleotides, and they are more likely to bind nonspecifically to proteins. Nevertheless, PS oligonucleotides have been shown to efficiently inhibit gene expression [28]. One concern with PS oligonucleotides (and all chemically modified oligonucleotides) is their potential for toxicity. The metabolic byproducts of PS oligonucleotides are not native to the cell. If these were incorporated into cellular DNA, they could cause mutations [29]. Studies have also shown that PS oligonucleotides activate complement, resulting in immunological complications for *in vivo* applications [30]. These complications and issues of affinity for the RNA, RNase H activity, stability, and cellular uptake have led to investigation

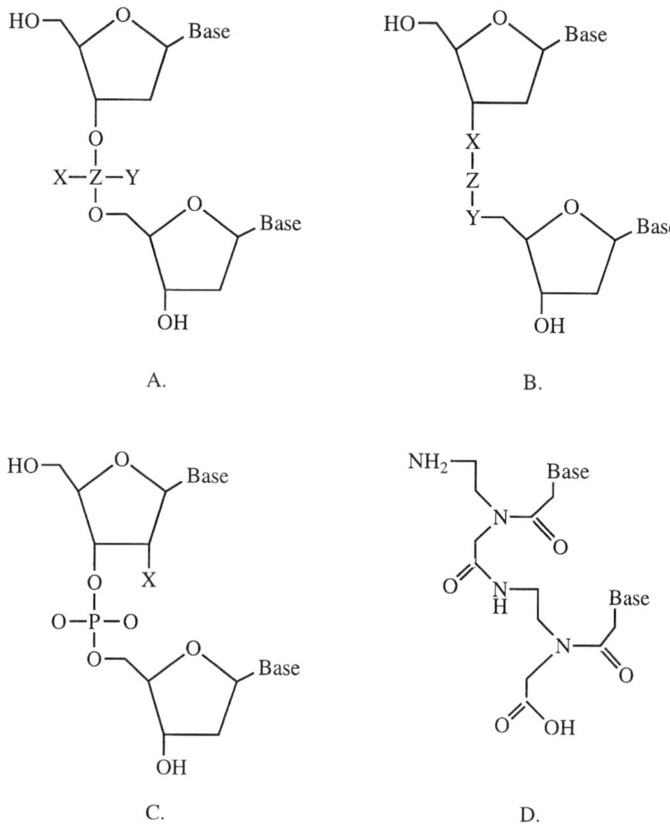

FIGURE 10.3 Analogs of natural oligonucleotides. (A) Backbone modifications in which the phosphorus bridge atom is retained. (B) Backbone modifications in which the phosphorus bridge atom is replaced. (C) 2′ ribose modifications. (D) Peptide nucleic acids — the entire backbone is replaced with amino acids. See Table 10.1 for legend.

of other types of modifications (e.g., N3′-P5′ phosphoramidates) and chimeric oligonucleotides (see Fig. 10.3 and Table 10.1).

One particularly promising modification focuses on the 2′ position of the sugar ring. RNA:RNA hybrids are more stable than RNA:DNA hybrids, presumably due to the formation of A-form helices rather than the A:B-form of the heteroduplex. However, with a hydroxyl at the 2′ position, RNA is exceptionally labile to nucleases. Thus, to maintain the A-form geometry (the preferred geometry for RNA helices) and prevent degradation, replacement of the 2′-hydroxyl is being investigated. A study by Monia, et al. investigated both 2′-halide and 2′-O-alkyl substitutions, finding the affinity for the target ranked 2′-fluoro > 2′-O-methyl > 2′-O-propyl > 2′-O-pentyl > 2′-O-deoxy [31]. However, fully modified oligonucleotides were ineffective in inhibiting Ha-ras gene expression in HeLa cells unless the interior of the molecule contained at least 5 2′-deoxy nucleotides to initiate RNase H cleavage. Hence, chimeric oligonucleotides have begun to receive more attention as the next generation of antisense therapeutics.

Chimeric oligonucleotides take advantage of the properties of multiple modifications in the same molecule. Earlier chimeric oligonucleotides used end caps of nuclease resistant bases with unmodified central bases [32]. More recently, "mixed-backbone" oligonucleotides have been synthesized with both backbone end modifications (primarily PS) and central sugar modifications (primarily 2′-O-methoxy) [33]. These molecules have been shown to have improved target affinity over PS oligonucleotides while maintaining the ability to initiate RNase H cleavage of the RNA.

Investigators have also replaced the phosphodiester backbone entirely, as with peptide nucleic acids (PNAs) in which the entire ribose-phosphodiester backbone is replaced with a polyamide (see Fig. 10.3C).

TABLE 10.1   The Names and Key Characteristics of Several Oligonucleotide Analogs

Phosphorus Analogs (Fig. 10.3a)	RNase H Activation	Nuclease Resistance	Chiral Center	Affinity	Charge	X (Fig. 10.3)	Y (Fig. 10.3)	Z (Fig. 10.3)
phosphodiester (PO)	yes	no	no	= PO	negative	O^-	O	P
phosphorothioate (PS)	yes	yes	yes	< PO	negative	S^-	O	P
methylphosphonate (MP)	no	yes	yes	< PO	neutral	CH_3	O	P
phosphoramidate	no	yes	yes	< PO	neutral	NH-R	O	P
phosphorodithioate	yes	yes	no	< PO	negative	S^-	S	P
phosphoethyltriester	no	yes	yes	> PO	neutral	$O-C_2H_5$	O	P
phosphoroselenoate	yes	yes	yes	< PO	negative	Se^-	O	P
Nonphosphorus Analogs (Fig. 10.3b)								
formacetal	?	yes	no	< PO	neutral	O	O	CH_2
3′ thioformacetal	?	yes	no	> PO	neutral	S	O	CH_2
5′-N-carbamate	?	yes	no	< or > PO	neutral	O	NH	C=O
sulfonate	?	yes	no	< PO	neutral	O	CH_2	SO_2
sulfamate	?	yes	no	< PO	neutral	O	NH	SO_2
sulfoxide	?	yes	no	< PO	neutral	CH_2	CH_2	SO
sulfide	?	yes	no	< PO	neutral	CH_2	CH_2	S
2′ Modified analogs (Fig. 10.3c)								
fluoro	no	yes	N/A	>PO	N/A[1]	F		
methoxy	no	yes	N/A	>PO	N/A[1]	$O-CH_3$		
propoxy	no	yes	N/A	>PO	N/A[1]	$O-(CH_2)_2CH_3$		
pentoxy	no	yes	N/A	>PO	N/A[1]	$O-(CH)_4CH_3$		
O-allyl	no	yes	N/A	>PO	N/A[1]	$O-CH=CH_2$		
methoxyethoxy	no	yes	N/A	>PO	N/A[1]	$O-(CH_2)_2-O-CH_3$		
α-analogs	no	yes	N/A[1]	< PO	N/A[1]	N/A[2]	N/A[2]	N/A[2]
Peptide nucleic acids (Fig. 10.3d)	no	yes	no	> PO	positive[3]	N/A[4]	N/A[4]	N/A[4]

? = unknown.

[1] Chirality and charge depends on backbone structure used.

[2] Structure not drawn; The bond between the sugar and base (an *N*-glycosidic bond) of α-analogs have the reverse orientation (α-configuration) from natural (β-configuration) oligonucleotides.

[3] Typically, the C-terminus is covalently linked to a positively charged lysine residue, giving the PNA a positive charge.

[4] See Fig. 10.3D for chemical structure.

X, Y, Z: these columns reference Fig. 10.3; Replace the designated letter in Fig. 10.3 with the molecule indicated in the Table I to determine the chemical structure of the oligo.

PNAs, with a covalently bound terminal lysine residue to prevent self-aggregation, have been shown to preferentially bind to complementary sequences and inhibit gene expression *in vitro* [34, 35]. PNAs have an unusually high affinity for their complementary target sequences, possibly because they can form a triple helix with homopurine stretches of RNA (PNA:RNA-PNA) [36]. Duplex forming PNAs have shown inhibition of translation when targeted to the mRNA start codon, while triple helix formation is required to inhibit protein elongation [37]. The ability of PNAs to invade the duplex of DNA and form a stable triple helix structure remains an issue for their *in vivo* application.

Other modifications to the ribose sugars (e.g., α-anomeric) and modifications to the nucleoside bases (e.g., 5-methyl or 5-bromo-2′-deoxycytidine and 7′-deazaguanosine and 7′-deazaadenosine oligonucleotides) have also been examined [10]. Oligonucleotides covalently linked to active groups (e.g., intercalators such as acridine, photoactivated crosslinking agents such as psoralens and chelating agents such as EDTA-Fe) are also being actively investigated as potential antisense molecules [9]. Whether for clinical application or for research/diagnostic tools, studies with modified oligonucleotides will provide valuable information regarding the mechanistic steps of antisense oligonucleotide activity. These future generations of antisense "oligonucleotides" may bear little structural resemblance to natural oligonucleotides, but their inhibition of gene expression will still rely on sequence specific base pairing. The reader is directed to several reviews for a comprehensive treatment of all classes of chemically modified oligonucleotides [9, 38-40].

10.4 Oligonucleotide Synthesis and Purification

The growth of the field of antisense therapeutics has also resulted in a need for increased production of oligonucleotides for *in vitro* and *in vivo* applications. As a result, new synthesis procedures are being explored to increase batch size and synthesis rate. Large-scale automated synthesizers have been developed, with which batches of 100 mmol have been produced [http://www.hybridon.com]. This is a significant improvement given that a large batch on a standard synthesizer is 1 μmol.

Standard procedures begin with the 3′ base of the oligonucleotide attached to a solid controlled pore glass (CPG) support, although other supports have also been proposed [41]. The process then continues with deprotection, monomer introduction, activated coupling, oxidation, and capping in a cycle for the addition of each oligonucleotide. Depending on the type of monomer being used (e.g., β-cyanoethyl phosphoramidite or hydrogen phosphonate), sulfurization to generate PS oligonucleotides occurs either during the synthesis cycling or following the termination of synthesis. Given the imperfections in the synthesis procedures, the resulting oligonucleotide product must then be purified to remove failure products.

Oligonucleotides are typically purified using reverse-phase, high-performance liquid chromatography (RP-HPLC). During synthesis, a hydrophobic dimethoxy-trityl (DMT) group is removed from the phosphoramidite to allow the next nucleoside to attach. Leaving this group on the final nucleoside of the synthesis increases the hydrophobicity of the full, correct oligo, increasing its retention time on an RP-HPLC column. The eluent is then fractionated to collect only the full sequence product.

However, PS oligonucleotides, unlike PO oligonucleotides, are chiral molecules, i.e., molecules that are not identical to their mirror image. Current synthesis techniques do not control the orientation of each internucleotide bond, and the resultant oligonucleotides are therefore a mixture of $2^{(n-1)}$ diastereomers (where n = the number of bases in the complete oligo). The resulting sequences elute with similar retention times on RP-HPLC as all of the complete sequence diastereomers should have the hydrophobic DMT group attached. However, analysis following removal of the DMT group can be used to confirm or deny the presence of multiple diastereomers under higher resolution conditions. Capillary gel electrophoresis and anion-exchange HPLC are being examined as ways to enhance the selectivity of purification of PS oligonucleotides. Additional details on the synthesis and purification of oligonucleotides are available for the interested reader [42-44].

10.5 Specificity of Oligonucleotides

The ability to block the expression of a single gene without undesired side effects (specificity) is the major advantage, in principle, of antisense based strategies. This specificity is primarily determined by the length (i.e., the number of bases) of the oligonucleotide. Experimental and theoretical data suggest that there is an optimum length at which specific inhibition of gene expression is maximized and nonspecific effects are minimized [13, 14].

Affinity and specificity limit the minimum effective length of oligonucleotides. Oligonucleotides that are too short do not inhibit gene expression, because they do not bind with sufficient affinity to their substrates. The shortest oligonucleotide reported to affect gene expression in mammalian cells was 7 bases in length (7-mer) [45]. This oligonucleotide was highly modified, containing 5-(1-propynyl) cytosines and uracils as well as a phosphorothioate backbone. An oligonucleotide that is too short is less likely to represent a unique sequence in a given cell's genome and thus more likely to bind to a non-targeted RNA and inhibit its expression. Woolf et al. estimated the minimum sequence length that will be statistically unique in a given pool of mRNAs [13]. Since each position in a given sequence can be occupied by any of four nucleotides (A, C, G or U), the total number of different possible sequences of length N bases is 4^N. Letting R equal the total number of bases in a given mRNA pool and assuming that it is a random and equal mixture of the four nucleotides, then the frequency (F) of occurrence in that pool of a sequence of length N is given by:

$$F = \frac{R}{4^N} \cdot$$ (10.1)

For a typical human cell, which contains approximately 10^4 unique mRNA species whose average length is 2000 bases, R is approximately equal to 2×10^7. Therefore, for a sequence to be unique ($F < 1$), N must be greater than or equal to 13 bases [13]. The minimum oligonucleotide length will also be constrained by the minimum binding affinity to form a stable complex.

However, oligonucleotides cannot be made arbitrarily long, because longer oligonucleotides are more likely to contain internal sequences complementary to nontargeted RNA molecules. This has also been expressed mathematically [13]. The expected number of complementary sites (S) of length L for an oligonucleotide of length N in an mRNA pool with R bases is given by:

$$S = \frac{\left[\left(N - L + 1 \right) \times R \right]}{4^L} \cdot$$ (10.2)

For example, an 18-mer (N = 18) has 6 internal 13-mers. Since a 13-mer is expected to occur 0.3 times in a mRNA pool containing 2×10^7 bases, the 18-mer is expected to match 1.8 (i.e., 6×0.3) 13-mers in the mRNA pool. Equation (10.2) also gives this result (N = 18, L = 13, and R = 2×10^7; therefore, S = 1.8).

Woolf et al. have demonstrated that significant degradation of nontargeted mRNAs can occur [13]. They compared the effectiveness of three different 25-mers in suppressing the expression of fibronectin mRNA in *Xenopus* oocytes. Nearly 80% of the fibronectin mRNA was degraded after the oocytes were microinjected with a 25-mer in which all 25 of its bases were complementary to the mRNA. However, when the oocytes were microinjected with 25-mers that had only 17 or 14 complementary bases flanked by random sequences, greater than 30% of their fibronectin mRNA was still degraded. They also showed that a single mismatch in an oligonucleotide did not completely eliminate its antisense effect. Over 40% of the target mRNA was degraded when oocytes were treated with a 13-mer with one internal mismatch, though the mismatch left a 9 base consecutive complementary sequence that showed nearly the same activity as a 13-mer with 10 complementary bases in succession.

Although these studies were conducted at lower temperatures and therefore under less stringent hybridization conditions than those found in mammalian cells, they clearly showed that complementary oligonucleotides flanked by unrelated sequences and even oligonucleotides with mismatched sequences can lead to significant degradation of RNA. The possibility of undesired inhibition of partially complementary sequences must be considered when designing and testing any antisense oligonucleotide. Hence, a useful heuristic is that an oligonucleotide should be long enough to be unique and possess high affinity for its target but short enough to minimize side effects due to degradation of nontargeted mRNAs.

The efforts to sequence the human genome have begun to provide more information about the actual specificity of oligonucleotides. It is possible to scan for sequences against all of the genomic information available using resources such as BLAST at the National Center for Biotechnology Information [46; http://www.ncbi.nlm.nih.gov]. The results will determine the uniqueness of the oligonucleotide within the known database sequences. As the completeness of the databases grows, computational comparisons of target RNA sequences against the database will provide more reliable assessments of the uniqueness of these target sequences within the genome.

10.6 Oligonucleotide Delivery

In many cases, the concentrations of antisense oligonucleotides required to achieve a biological effect are currently too large to be of therapeutic value. Achieving the necessary concentration in the vicinity of the target cells is thought to be limited at least in part by charge repulsion between the negatively charged oligonucleotide and the negatively charged cell surface. Four major methods are currently being investigated to overcome this and other extracellular transport barriers: (1) chemical modification of the oligonucleotide to increase its hydrophobicity; (2) conjugation of the oligonucleotide to a polycation; (3) conjugation of the oligonucleotide to a ligand specific for a cellular receptor; and (4) encapsulation of the oligonucleotide in a liposome.

Increasing the hydrophobicity of oligonucleotides was first accomplished by synthesis of methylphosphonate backbones. Despite their neutral backbone charge, cellular uptake of MP oligonucleotides was shown to be very inefficient, perhaps due in part to their inability to escape lysosomal degradation and gain entry into the cytoplasm. The hydrophobicity of an oligonucleotide can also be increased by conjugation with hydrophobic moieties such as cholesterol derivatives (chol-oligonucleotide). The increased hydrophobicity of chol-oligonucleotides reportedly improves their association with cell membranes and their internalization by cells [47]. Cellular accumulation of chol-oligonucleotides after 2 hours has been shown to occur when no accumulation was seen with unconjugated oligonucleotides [48]. The mechanism by which chol-oligonucleotides are more easily and rapidly taken up by cells has not been elucidated. Krieg [49] showed that chol-oligonucleotides with cholesterol at the 5′ end could be bound by low density lipoprotein (LDL) and that this markedly increased the association of the oligonucleotide with the cell membrane and its internalization *in vitro*. LDL-associated chol-oligonucleotides were effective at 8-fold lower concentration than PO oligonucleotide controls.

Oligonucleotides conjugated with polycations such as poly-L-lysine (PLL) also have improved cellular uptake. Covalent conjugates can be constructed by coupling the 3′ end of the oligonucleotide to the epsilon amino groups of the lysine residues [50]. Oligo-PLL conjugates complementary to the translation initiation site of the *tat* gene (a key HIV regulatory protein) protected cells from HIV-1 infection with concentrations 100-fold lower than nonconjugated oligonucleotides [51]. Low concentrations of oligo-PLL conjugates (100 nM), in which the 15-mer oligonucleotide was complementary to the initiation region of the viral N protein mRNA, also inhibited vesicular stomatitis virus (VSV) infection [50].

Internalization of oligonucleotides has also been improved by conjugation to ligands. One such ligand, transferrin, is internalized by receptor mediated endocytosis. Mammalian cells acquire iron carried by transferrin. Oligo-PLL complexes have been conjugated to transferrin to take advantage of this pathway. For example, an 18-mer complementary to *c-myb* mRNA (an oncogene that is responsible for the hyperproliferation of some leukemia cells) was complexed with a transferrin PLL conjugate and rapidly internalized by human leukemia (HL-60) cells [52]. The expression of *c-myb* was greatly reduced and the uncontrolled proliferation of these cells inhibited. Because an oligo-PLL conjugate (without transferrin) was not tested, it is not clear whether the improved antisense effects were due to the PLL moiety, the transferrin moiety, or a combination thereof.

Oligo-ligand conjugates have also been used to target oligonucleotides to cells with specific cellular receptors, such as the asialoglycoprotein (ASGP) receptor, which is expressed uniquely by hepatocytes. These receptors bind and internalize ASGPs, serum glycoproteins that have exposed galactose residues at the termini of their glycosylation chains. Oligonucleotides complexed to ASGP are rapidly internalized by hepatocytes. A number of researchers have used ASGP conjugated to cationic PLL. When the ASGP:PLL is conjugated to oligonucleotides, their cellular uptake and antisense effectiveness *in vitro* is increased [53-55]. However, toxicity of the PLL moiety and instability of the noncovalent complexes between ASGP-PLL and oligonucleotides have limited their *in vivo* applicability [56]. One possible solution is the direct conjugation of the oligonucleotides to ASGP via a cleavable disulfide linkage [57].

Another promising means to deliver oligonucleotides is lipofection. Oligonucleotides are mixed with cationic lipids that condense around the negatively charged oligonucleotide forming a lipid vesicle (liposome). The positively charged lipids reduce the electrostatic repulsion between the negatively-charged oligonucleotide and the similarly charged cell surface. Bennett et al. made liposomes using DOTMA {*N*-{1-(2,3-dioleyloxy)propyl)}-*N,N,N*-trimethylammonium chloride} as the cationic lipid and an oligonucleotide complementary to the translation initiation codon of human intracellular adhesion molecule 1 (ICAM-1), increasing the potency of the oligonucleotide by >1000-fold [12]. A recent study has shown that liposomally encapsulated oligonucleotides at 0.01 nM were as active as free oligonucleotides at 1.5 μM at inhibiting HIV-1 replication in chronically infected cells [58]. It was found that the oligonucleotides separate from cationic liposomes following cellular internalization [59]. Cationic lipids subsequently accumulate in cytoplasmic structures and the plasma membrane, suggesting that they have fused with the endosomal vesicles and are being recirculated throughout the secretory pathways [59-61]. The increase in oligonucleotide activity suggests that cellular uptake may be a significant limitation in the effectiveness of antisense oligonucleotides. *In vitro* studies using electroporation, streptolysin O or α-toxin permeabilization, and particle bombardment to enhance cellular uptake of nucleic acids support this hypothesis [62-65].

Efforts have been made to use antibodies to target liposomes containing oligonucleotides to specific cell types. Protein A, which binds to many IgG antibodies, is covalently bound to the liposome to form a liposome-oligo-protein A complex. An antibody specific for the target cell's surface is bound to the protein A, and these complexes are incubated with the target cells. Leonetti [66] synthesized such a complex to target the mouse major histocompatibility complex H-2K molecule on mouse L929 cells. The complex contained a 15-mer complementary to the 5′ end of mRNA encoding the N protein of VSV. The complexes inhibited viral replication by more than 95%. Unconjugated liposomes, and liposomes conjugated to antibodies specific for a nonexpressed antigen, had no effect.

10.7 Potential Applications of Antisense Oligonucleotides

The potential applications for antisense oligonucleotides are limited only by the genetic information available. Antisense oligonucleotides can be developed against any target in which the inhibition of protein production or the inhibition of RNA processing yields the therapeutic result. Currently, clinical trials are underway using antisense oligonucleotides to treat rheumatoid arthritis, psoriasis, renal transplant rejection, and inflammatory bowel disease (Crohn's disease) [http://www.phrma.org]. However, the primary targets remain refractory viral diseases and cancers for which the necessary target genetic information is typically available.

Viral Diseases

One of the many potential therapeutic applications of antisense technology is the treatment of infectious viral diseases. The use of antisense oligonucleotides as antiviral agents is particularly promising because viral nucleic acid sequences are unique to the infected cell and are not found in normal healthy cells. The goal of antiviral therapy is to block the expression of key viral proteins that are vital to the life cycle of the virus. This has been achieved *in vitro* with several viruses including HIV, HSV, influenza and human papilloma virus [10]. The *in vitro* work with HIV is representative of other viruses and will be highlighted here.

Human Immunodeficiency Virus

Retroviruses, and HIV in particular, have high rates of mutation and genetic recombination. The effectiveness of many anti-HIV drugs has been severely reduced because drug resistant viral strains often arise after prolonged drug treatment. This is especially relevant to strategies that use an antisense approach that relies on a specific nucleotide sequence. One strategy to inhibit HIV replication is to target conserved sequences in key regulatory proteins such as *tat* and *rev*. Part of the *rev* sequence is highly conserved, with the known 16 isolates of HIV differing by at most one base pair in this conserved region [67]. The importance of targeting a highly conserved region as is found in *rev* was demonstrated by Lisziewicz, who investigated the efficacy of antisense oligonucleotides directed against various viral proteins [68]. Although several 28-mers initially inhibited viral replication, resistant mutant viruses developed after 25 days to all of the oligonucleotides with the exception of those directed at the highly conserved *rev* regions.

A second concern in treating HIV and other viruses is that viral replication can restart after antisense treatment is stopped [67]. Can oligonucleotides be continuously administered to prevent viral replication without unwanted side effects? These issues and others common to all antisense based therapies (oligonucleotide stability, specificity, affinity, and delivery) must be addressed prior to the successful implementation of antisense-based HIV therapies.

Cancer

In principle, antisense technology can be used against cancer, but the target is more challenging. Oncogenes are typically genes that play a vital regulatory role in the growth of a cell, the mutation or inappropriate expression of which can result in the development of cancer. In the case of mutation, it is often difficult to distinguish an oncogene from its normal counterpart, because they may differ by as little as one base. Thus, attempts to inhibit oncogene expression might block the expression of the normal gene in non-cancerous cells and cause cytotoxic effects. Despite these challenges, steady progress has been made in the development of effective antisense oligonucleotides that have inhibited many types of oncogenes including *bcr-abl, c-myc, c-myb, c-ras,* Ha-*ras, neu/erbB2* and *bFGF* [11, 25, 61, 69, 70].

Studies targeting *ras* oncogenes are encouraging and are representative of progress against other classes of oncogenes. Chang [71] targeted a *ras p21* gene point mutation with an antisense oligonucleotide. Only the expression of the mutated genes was inhibited, suggesting that it is possible to selectively inhibit an oncogene that differs by only a single base from the normal gene. Using a 17-mer PS oligo, it was shown that ~98% inhibition of Ha-*ras* expression could be achieved at 1 μM concentration [25].

Successes in inhibiting oncogene expression *in vitro* are encouraging, but many problems remain to be solved before antisense oligonucleotides are therapeutically useful for cancer patients. For example, finding a suitable oncogene to target is difficult. Even if a genetic defect is common to all cells of a given cancer, there is no guarantee that inhibition of that oncogene will halt cancer cell proliferation [70]. Even if cell growth is inhibited, tumor growth may restart after antisense treatment is stopped. Oligonucleotides are needed that induce cancer cell terminal differentiation or death [70]. In order to avoid toxic side effects by inhibiting gene expression of regulated oncogenes in normal tissues, oligonucleotides that specifically target cancer cells are required. Even if appropriate antisense oligonucleotides are developed, they will still be ineffective against solid tumors unless they can reach the interior of the tumor at biologically effective concentrations.

10.8 *In Vivo* Pharmacology

The first *in vivo* antisense studies were designed to test the biodistribution and toxicity of antisense oligonucleotides. These studies demonstrated that PO oligonucleotides are rapidly excreted from the body with PS oligonucleotides being retained longer. One study in mice noted that completely modified PS oligonucleotides were retained significantly longer than a chimeric 20-mer that contained 15 phosphodiester bonds and only 4 phosphorothioate bonds [72]. Only 30% of the PS oligonucleotide was excreted in the urine after 24 hours, whereas 75% of the chimeric oligonucleotide was excreted after only 12 hours. A similar study demonstrated that PS oligonucleotides were retained in body tissues of adult male rats with a half-life of 20 to 40 hours [73].

These studies suggested that antisense oligonucleotides may need to be administered repeatedly to maintain a therapeutic effect for chronic disorders. At least with PS oligonucleotides, this has proven to be problematic. Continuous or repeated administration of PS oligonucleotides results in immunologic complement activation. It has been shown that infusion of PS oligonucleotides to rhesus monkeys results in almost immediate (within 10 min.) decreases in white blood cell count and the fraction of neutrophils among the white cells [30]. At 40 min. post-infusion, the neutrophil counts had increased to higher than baseline levels. A recent study indicated that a plasma concentration of >50 μg/ml results in complement activation [74]. These difficulties have further hastened the development of novel oligonucleotide chemistries and chimeric oligonucleotides for which the immunologic properties are expected to be improved due to the possible reduction in the use of PS backbone linkages [33, 75].

The targeted delivery of oligonucleotides has been proposed as a way to improve the effectiveness of systemically administered antisense oligonucleotides by oligonucleotide concentration in the desired cells and minimizing nonspecific side effects. Targeting oligonucleotides to a specific organ has been demonstrated. As previously discussed, a soluble DNA carrier system has been developed to specifically target hepatocytes [76]. This system was used to target a 67-mer PO oligonucleotide (complementary to the 5′ end of rat serum albumin mRNA) to the liver [56]. Following tail vein injection into rats, the complex rapidly and preferentially accumulated in the liver, but the efficiency of this targeting method was limited by the rapid dissociation of the oligonucleotide from the ASGP:PLL complex (30% dissociated within 7 minutes).

10.9 Animal Models

The effectiveness of antisense oligonucleotides at inhibiting gene expression *in vivo* has been demonstrated in several animal models. Immunodeficient mice, bearing a human myeloid leukemia cell line, were continuously infused with 100 μg/day of a 24-mer PS oligonucleotide complementary to c-*myb* mRNA [77]. Mice that received the antisense oligonucleotide survived on average 8.5 times longer than control mice that received no treatment and 5.7 times longer than sense-treated controls. The treated animals also had a significantly lower tumor burden in the ovaries and brain.

Antisense oligonucleotides have also inhibited the growth of solid tumors *in vivo*. Immunodeficient mice bearing a fibrosarcoma or melanoma were injected subcutaneously twice weekly with 1.4 mg of PS oligonucleotides complementary to the 5′ end of the p65 subunit of the mRNA for the NF-κB transcription factor (NF-κB activates a wide variety of genes and is believed to be important in cell adhesion and tumor cell metastasis) [78]. Greater than 70% of antisense treated mice exhibited a marked reduction in their tumor size. Administration of control oligonucleotides complementary to GAPDH and *jun-D* had no effect on tumor size.

The capability of antisense oligonucleotides to inhibit viral replication *in vivo* has also been studied. Fourteen one-day-old ducklings were infected with duck hepatitis B virus (DHBV), which is closely related to the strain of hepatitis B virus that is a major cause of chronic liver disease and cancer in humans [79]. Two weeks later, the ducks were injected daily, for 10 consecutive days, with 20 μg/gm body weight of an 18-mer PS oligonucleotide complementary to the start site of the Pre-S region of the DHBV genome. The antisense oligonucleotides blocked viral gene expression and eliminated the appearance of viral antigens in the serum and liver of all treated ducks. No toxic effects due to the oligonucleotides were noted. Unfortunately, residual amounts of DNA precursors of viral transcripts were detected in the nuclei of liver cells, which resulted in a slow restart of viral replication after antisense treatment was stopped. Further studies are needed to determine if prolonged treatment with antisense will eliminate this residual viral DNA.

Though many of the oligonucleotides under investigation target viruses or cancer, other refractory targets are also being investigated. Antisense oligonucleotides have been evaluated in the treatment of restenosis, a narrowing of an artery following corrective surgery caused by mitogen-induced proliferation of vascular smooth muscle cells (SMC). Several studies have tested the ability of antisense oligonucleotides to inhibit genes whose expression is important in SMC proliferation, including c-*myc*, c-*myb*, *cdc2* and *cdk2* [80-82]. In one study with an antisense therapeutic complementary to c-*myc* mRNA, the formation of a thick intimal layer (neointima) in treated areas was significantly reduced in antisense treated rats when compared to control rats with maximum c-*myc* expression reduced by 75% relative to controls [82]. Dextran sodium sulfate-induced colitis in Swiss-Webster mice (as measured by disease activity index, DAI) was reduced by ~64% following 5 days of daily subcutaneous administration of an antisense oligonucleotide complementary to ICAM-1 at 0.3 mg/kg/day [83]. Interestingly, in this study, higher oligonucleotide concentrations (>5 mg/kg/day) were ineffective in reducing inflammation.

their cellular permeability, the conjugation of oligonucleotides to specific ligands to utilize more efficient receptor-mediated internalization pathways, and the use of antibody conjugated liposomes to deliver oligonucleotides to specific cells.

The potential applications of antisense oligonucleotides to the treatment of disease are vast. Antisense-based therapies are under development for the treatment of infectious diseases such as CMV, HIV, herpes simplex virus, influenza, hepatitis and human papilloma virus as well as the treatment of complex genetic disorders like cancer. Animal models are being used to determine if (1) antisense oligonucleotides can be delivered to target cells at high enough concentrations to be effective, (2) repeated treatments with oligonucleotides are toxic or elicit an immune response, and (3) antisense oligonucleotides directed against a single gene can be effective against complex genetic diseases such as cancer. Multiple clinical trials against viral, cancer, and other targets are now in progress with one drug having been approved for the clinic. Improvements in our understanding of the mechanisms of antisense inhibition, the pharmacology of antisense oligonucleotides *in vivo* and the development of chemically modified oligo-nucleotides with high affinity, specificity, and stability are needed to realize the clinical potential of antisense-based strategies for the treatment of a wide variety of diseases.

Defining Terms

Antisense: Any DNA or RNA molecule whose sequence is complementary to the RNA transcribed from a target gene.

Chiral: A molecule whose configuration is not identical with its mirror image.

Complementary: A nucleic acid sequence is complementary to another if it is able to form a perfectly hydrogen-bonded duplex with it, according to the Watson-Crick rules of base pairing (A opposite U or T, G opposite C).

Diastereomer: Optically active isomers that are not enantiomorphs (mirror images).

Exonuclease: An enzyme that catalyzes the release of one nucleotide at a time, serially, from one end of a polynucleotide.

In vitro: In an artificial environment, referring to a process or reaction occurring therein, as in a test tube or culture dish.

In vivo: In the living body, referring to a process or reaction occurring therein.

Lipofection: Delivery of therapeutic drugs (antisense oligonucleotides) to cells using cationic liposomes.

Lipophilic: Capable of being dissolved in lipids (organic molecules that are the major structural elements of biomembranes).

Liposome: A spherical particle of lipid substance suspended in an aqueous medium.

Plasmid: A small, circular extrachromosomal DNA molecule capable of independent replication in a host cell.

Receptor mediated endocytosis: The selective uptake of extracellular proteins, oligonucleotides, and small particles, usually into clathrin-coated pits, following their binding to cell surface receptor proteins.

Restenosis: A narrowing of an artery or heart valve following corrective surgery on it.

RNase H: An enzyme that specifically recognizes RNA:DNA duplexes and cleaves the RNA portion, leaving the DNA portion intact.

References

1. Zamecnik, P. C. and Stephenson, M. L. 1978. Inhibition of Rous sarcoma virus replication and cell transformation by a specific oligodeoxynucleotide, *Proc. Natl. Acad. Sci. USA*, 75, 280.
2. Tomizawa, J. and Itoh, T. 1981. Plasmid ColE1 incompatibility determined by interaction of RNA I with primer transcript, *Proc. Natl. Acad. Sci. USA*, 78, 6096.
3. Tomizawa, J., Itoh, T., Selzer, G., and Som, T. 1981. Inhibition of ColE1 RNA primer formation by a plasmid-specified small RNA, *Proc. Natl. Acad. Sci. USA*, 78, 1421.

10.10   Clinical Trials

The FDA recently approved the first antisense therapeutic, fomivirsen, a treatment for cytomegalovirus (CMV) retinitis in AIDS patients [http://www.isip.com]. CMV infection is fairly common (~50% of people over 35 are seropositive); however, it results in complications only in immunocompromised individuals. The oligonucleotide is complementary to the immediate-early viral RNA, a region encoding proteins that regulate virus gene expression [84]. The activity of the oligonucleotide therapeutic is potentiated by traditional drugs such as ganciclovir. Additional studies suggested, however, that the oligonucleotide acts not only via an antisense mechanism, though complementarity to the target was required for maximal inhibitory capacity [85].

Clinical trials on many other antisense oligonucleotides have advanced to Phase II. However, complications associated with antisense oligonucleotides have also been found. An antisense compound targeted to the gag region of HIV was pulled from clinical trials due to the resulting decreases in platelet counts in 30% of the patients after 10 days of daily administration [http://www.hybridon.com]. The immunologic complications are likely related to the activation of complement discussed earlier. Perhaps, fomivirsen avoids these complications due to its local administration by intravitreal injection rather than systemic infusion. Table 10.2 lists clinical trials that are currently ongoing and the disease targets. Published results from these trials are not yet available.

10.11   Summary

Antisense oligonucleotides have the potential to selectively inhibit the expression of any gene with a known sequence. Although there have been some remarkable successes, realizing this potential is proving difficult because of problems with oligonucleotide stability, specificity, affinity, and delivery. Natural oligonucleotides contain a phosphodiester backbone that is highly susceptible to nuclease degradation. Chemically modified oligonucleotides (PS and chimeric oligonucleotides) have been synthesized that are more stable. However, the increased stability of PS oligonucleotides comes at a cost, in particular the activation of complement. Another issue for chemically modified oligonucleotides is their large scale production and purification. Alternative chemistries are needed that are more biologically active and more amenable to large scale production. Targeted delivery of antisense oligonucleotides is needed to minimize side effects and maximize oligonucleotide concentration in target cells. Attempts to improve delivery include chemical modification of the oligonucleotides to increase

TABLE 10.2   A Sample of Current Clinical Trials of Antisense Oligodeoxynucleotide Therapeutics

Compound	Type of Oligo	Disease Target	Gene Target	Phase
ISIS13312	2′-modified — 2nd Gen.	CMV retinitis	N/A	I/II
ISIS2302	PS	Crohn's disease	ICAM-1	II/Pivotal
		Psoriasis	ICAM-1	II
		Renal transplant rejection	ICAM-1	II
		Rheumatoid arthritis	ICAM-1	II
		Ulcerative colitis	ICAM-1	II
G3139	N/A	Cancer	N/A	I
ISIS2503	PS	Cancer	Ha-ras	I
ISIS3521	PS	Cancer	PKC-α	II
ISIS5132	PS	Cancer	c-raf	II
LR-3001	PS	Chromic myelogenous leukemia	bcr-abl	I
LR-3280	PS	Restenosis	myc	II

N/A — Not available
Potential applications of antisense technology include the treatment of cancer, infectious diseases, and inflammatory diseases.

4. Izant, J. G. and Weintraub, H. 1985. Constitutive and conditional suppression of exogenous and endogenous genes by anti-sense RNA, *Science*, 229, 345.

5. Stull, R. A. and Szoka, F. C., Jr. 1995. Antigene, ribozyme and aptamer nucleic acid drugs: progress and prospects, *Pharm Res*, 12, 465.

6. Hélène, C., Giovannangeli, C., Guieysse-Peugeot, A. L., and Praseuth, D. 1997. Sequence-specific control of gene expression by antigene and clamp oligonucleotides, *Ciba Found. Symp.*, 209, 94.

7. Gewirtz, A. M., Sokol, D. L., and Ratajczak, M. Z. 1998. Nucleic acid therapeutics: State of the art and future prospects, *Blood*, 92, 712.

8. Rossi, J. J. 1997. Therapeutic applications of catalytic antisense RNAs (ribozymes), *Ciba Found. Symp.*, 209, 195.

9. Hélène, C. and Toulme, J. J. 1990. Specific regulation of gene expression by antisense, sense and antigene nucleic acids, *Biochem. Biophys. Acta.*, 1049, 99.

10. Milligan, J. F., Matteucci, M. D., and Martin, J. C. 1993. Current concepts in antisense drug design, *J. Med. Chem.*, 36, 1923.

11. Nagel, K. M., Holstad, S. G., and Isenberg, K. E. 1993. Oligonucleotide pharmacotherapy: An antigene strategy, *Pharmacotherapy*, 13, 177.

12. Bennett, C. F., Condon, T. P., Grimm, S., Chan, H., and Chiang, M. Y. 1994. Inhibition of endothelial cell adhesion molecule expression with antisense oligonucleotides, *J. Immunol.*, 152, 3530.

13. Woolf, T. M., Melton, D. A., and Jennings, C. G., 1992. Specificity of antisense oligonucleotides in vivo, *Proc. Natl. Acad. Sci. USA*, 89, 7305.

14. Herschlag, D. 1991. Implications of ribozyme kinetics for targeting the cleavage of specific RNA molecules *in vivo*: more isn't always better, *Proc. Natl. Acad. Sci. USA*, 88, 6921.

15. Stull, R. A., Zon, G., and Szoka, F. C., Jr. 1996. An *in vitro* messenger RNA binding assay as a tool for identifying hybridization-competent antisense oligonucleotides, *Antisense Nucleic Acid Drug Dev.*, 6, 221.

16. Milner, N., Mir, K. U., and Southern, E. M. 1997. Selecting effective antisense reagents on combinatorial oligonucleotide arrays, *Nat. Biotechnol.*, 15, 537.

17. Jaroszewski, J. W., Syi, J. L., Ghosh, M., Ghosh, K., and Cohen, J. S. 1993. Targeting of antisense DNA: comparison of activity of anti-rabbit beta- globin oligodeoxyribonucleoside phosphorothioates with computer predictions of mRNA folding, *Antisense Res. Dev.*, 3, 339.

18. Sczakiel, G., Homann, M., and Rittner, K. 1993. Computer-aided search for effective antisense RNA target sequences of the human immunodeficiency virus type 1, *Antisense Res. Dev.*, 3, 45.

19. Lima, W. F., Monia, B. P., Ecker, D. J., and Freier, S. M. 1992. Implication of RNA structure on antisense oligonucleotide hybridization kinetics, *Biochemistry*, 31, 12055.

20. Monia, B. P., Johnston, J. F., Geiger, T., Muller, M., and Fabbro, D. 1996. Antitumor activity of a phosphorothioate antisense oligodeoxynucleotide targeted against C-raf kinase, *Nat. Med.*, 2, 668.

21. Goodchild, J., Carroll, E. D., and Greenberg, J. R. 1988. Inhibition of rabbit beta-globin synthesis by complementary oligonucleotides: identification of mRNA sites sensitive to inhibition, *Arch. Biochem. Biophys.*, 263, 401.

22. Wakita, T. and Wands, J. R. 1994. Specific inhibition of hepatitis C virus expression by antisense oligodeoxynucleotides. *In vitro* model for selection of target sequence, *J. Biol. Chem.*, 269, 14205.

23. Peyman, A., Helsberg, M., Kretzschmar, G., Mag, M., Grabley, S., and Uhlmann, E. 1995. Inhibition of viral growth by antisense oligonucleotides directed against the IE110 and the UL30 mRNA of herpes simplex virus type-1, *Biol. Chem. Hoppe Seyler*, 376, 195.

24. McKay, R. A., Cummins, L. L., Graham, M. J., Lesnik, E. A., Owens, S. R., Winniman, M., and Dean, N. M. 1996. Enhanced activity of an antisense oligonucleotide targeting murine protein kinase C-alpha by the incorporation of 2'-O-propyl modifications, *Nucleic Acids Res.*, 24, 411.

25. Monia, B. P., Johnston, J. F., Sasmor, H., and Cummins, L. L. 1996. Nuclease resistance and antisense activity of modified oligonucleotides targeted to Ha-ras, *J. Biol. Chem.*, 271, 14533.

26. Iribarren, A. M., Cicero, D. O., and Neuner, P. J. 1994. Resistance to degradation by nucleases of (2'S)-2'-deoxy-2'-C-methyloligonucleotides, novel potential antisense probes, *Antisense Res. Dev.*, 4, 95.

27. Loke, S. L., Stein, C. A., Zhang, X. H., Mori, K., Nakanishi, M., Subasinghe, C., Cohen, J. S., and Neckers, L. M. 1989. Characterization of oligonucleotide transport into living cells, *Proc. Natl. Acad. Sci. USA*, 86, 3474.

28. Stein, C. A., Subasinghe, C., Shinozuka, K., and Cohen, J. S. 1988. Physicochemical properties of phosphorothioate oligodeoxynucleotides, *Nucleic Acids Res.*, 16, 3209.

29. Neckers, L. and Whitesell, L. 1993. Antisense technology: biological utility and practical considerations, *Am. J. Physiol.*, 265, L1.

30. Galbraith, W. M., Hobson, W. C., Giclas, P. C., Schechter, P. J., and Agrawal, S. 1994. Complement activation and hemodynamic changes following intravenous administration of phosphorothioate oligonucleotides in the monkey, *Antisense Res. Dev.*, 4, 201.

31. Monia, B. P., Lesnik, E. A., Gonzalez, C., Lima, W. F., McGee, D., Guinosso, C. J., Kawasaki, A. M., Cook, P. D., and Freier, S. M. 1993. Evaluation of 2'-modified oligonucleotides containing 2'-deoxy gaps as antisense inhibitors of gene expression, *J. Biol. Chem.*, 268, 14514.

32. Pickering, J. G., Isner, J. M., Ford, C. M., Weir, L., Lazarovits, A., Rocnik, E. F., and Chow, L. H. 1996. Processing of chimeric antisense oligonucleotides by human vascular smooth muscle cells and human atherosclerotic plaque. Implications for antisense therapy of restenosis after angioplasty, *Circulation*, 93, 772.

33. Agrawal, S., Jiang, Z., Zhao, Q., Shaw, D., Cai, Q., Roskey, A., Channavajjala, L., Saxinger, C., and Zhang, R. 1997. Mixed-backbone oligonucleotides as second generation antisense oligonucleotides: *in vitro* and *in vivo* studies, *Proc. Natl. Acad. Sci. USA*, 94, 2620.

34. Hanvey, J. C., et al. 1992. Antisense and antigene properties of peptide nucleic acids, *Science*, 258, 1481.

35. Nielsen, P. E., Egholm, M., Berg, R. H., and Buchardt, O. 1993. Peptide nucleic acids (PNAs): potential antisense and anti-gene agents, *Anticancer Drug Des.*, 8, 53.

36. Egholm, M., Nielsen, P. E., Buchardt, O., and Berg, R. H. 1992. Recognition of guanine and adenine in DNA by cytosine and thymine containing peptide nucleic acids (PNA), *J. Am. Chem. Soc.*, 114, 9677.

37. Good, L. and Nielsen, P. E. 1997. Progress in developing PNA as a gene-targeted drug, *Antisense Nucleic Acid Drug Dev.*, 7, 431.

38. Uhlmann, E. and Peyman, A. 1990. Antisense oligonucleotides: A new therapeutic principle, *Chem. Rev.*, 90, 544.

39. Cook, P. D. 1991. Medicinal chemistry of antisense oligonucleotides — Future opportunities, *Anticancer Drug Des.*, 6, 585.

40. Matteucci, M. 1997. Oligonucleotide analogues: an overview, *Ciba Found. Symp.*, 209, 5.

41. Fearon, K. L., Hirschbein, B. L., Chiu, C. Y., Quijano, M. R., and Zon, G. 1997. Phosphorothioate oligodeoxynucleotides: large-scale synthesis and analysis, impurity characterization, and the effects of phosphorus stereochemistry, *Ciba Found. Symp.*, 209, 19.

42. Righetti, P. G. and Gelfi, C. 1997. Recent advances in capillary electrophoresis of DNA fragments and PCR products, *Biochem. Soc. Trans.*, 25, 267.

43. Righetti, P. G. and Gelfi, C. 1997. Recent advances in capillary electrophoresis of DNA fragments and PCR products in poly(n-substituted acrylamides), *Anal. Biochem.*, 244, 195.

44. Schlingensiepen, R., Brysch, W., and Schlingensiepen, K.-H. 1997. *Antisense — From technology to therapy*, Blackwell Wissenschaft Berlin, Vienna, Austria.

45. Wagner, R. W., Matteucci, M. D., Grant, D., Huang, T., and Froehler, B. C. 1996. Potent and selective inhibition of gene expression by an antisense heptanucleotide, *Nat. Biotechnol.*, 14, 840.

46. Altschul, S. F., Madden, T. L., Schaffer, A. A., Zhang, J., Zhang, Z., Miller, W., and Lipman, D. J. 1997. Gapped BLAST and PSI-BLAST: A new generation of protein database search programs, *Nucleic Acids Res.*, 25, 3389.

47. Boutorine, A. S. and Kostina, E. V. 1993. Reversible covalent attachment of cholesterol to oligodeoxyribonucleotides for studies of the mechanisms of their penetration into eucaryotic cells, *Biochimie*, 75, 35.

48. Alahari, S. K., Dean, N. M., Fisher, M. H., Delong, R., Manoharan, M., Tivel, K. L., and Juliano, R. L. 1996. Inhibition of expression of the multidrug resistance-associated P- glycoprotein of by phosphorothioate and 5′ cholesterol-conjugated phosphorothioate antisense oligonucleotides, *Mol. Pharmacol.*, 50, 808.

49. Krieg, A. M., Tonkinson, J., Matson, S., Zhao, Q., Saxon, M., Zhang, L. M., Bhanja, U., Yakubov, L., and Stein, C. A. 1993. Modification of antisense phosphodiester oligodeoxynucleotides by a 5′ cholesteryl moiety increases cellular association and improves efficacy, *Proc. Natl. Acad. Sci. USA*, 90, 1048, 1993.

50. Leonetti, J. P., Degols, G., Clarenc, J. P., Mechti, N., and Lebleu, B. 1993. Cell delivery and mechanisms of action of antisense oligonucleotides, *Prog. Nucleic Acid Res. Mol. Biol.*, 44, 143.

51. Degols, G., Leonetti, J. P., Benkirane, M., Devaux, C., and Lebleu, B. 1992. Poly(L-lysine)-conjugated oligonucleotides promote sequence-specific inhibition of acute HIV-1 infection, *Antisense Res. Dev.*, 2, 293.

52. Citro, G., Perrotti, D., Cucco, C., I, D. A., Sacchi, A., Zupi, G., and Calabretta, B. 1992. Inhibition of leukemia cell proliferation by receptor-mediated uptake of *c-myb* antisense oligodeoxynucleotides, *Proc. Natl. Acad. Sci. USA*, 89, 7031.

53. Bonfils, E., Depierreux, C., Midoux, P., Thuong, N. T., Monsigny, M., and Roche, A. C. 1992. Drug targeting: synthesis and endocytosis of oligonucleotide- neoglycoprotein conjugates, *Nucleic Acids Res.*, 20, 4621.

54. Wu, G. Y. and Wu, C. H. 1992. Specific inhibition of hepatitis B viral gene expression *in vitro* by targeted antisense oligonucleotides, *J. Biol. Chem.*, 267, 12436.

55. Roth, C. M., Reiken, S. R., Le Doux, J. M., Rajur, S. B., Lu, X.-M., Morgan, J. R., and Yarmush, M. L. 1997. Targeted antisense modulation of inflammatory cytokine receptors, *Biotechnol. Bioeng.*, 55, 72.

56. Lu, X. M., Fischman, A. J., Jyawook, S. L., Hendricks, K., Tompkins, R. G., and Yarmush, M. L. 1994. Antisense DNA delivery in vivo: Liver targeting by receptor-mediated uptake, *J. Nucl. Med.*, 35, 269.

57. Rajur, S. B., Roth, C. M., Morgan, J. R., and Yarmush, M. L. 1997. Covalent protein-oligonucleotide conjugates for efficient delivery of antisense molecules, *Bioconjugate Chem.*, 8, 935.

58. Lavigne, C. and Thierry, A. R. 1997. Enhanced antisense inhibition of human immunodeficiency virus type 1 in cell cultures by DLS delivery system, *Biochem. Biophys. Res. Commun.*, 237, 566.

59. Zelphati, O. and Szoka, F. C., Jr. 1996. Mechanism of oligonucleotide release from cationic liposomes, *Proc. Natl. Acad. Sci. USA*, 93, 11493.

60. Marcusson, E. G., Bhat, B., Manoharan, M., Bennett, C. F., and Dean, N. M. 1998. Phosphorothioate oligodeoxyribonucleotides dissociate from cationic lipids before entering the nucleus, *Nucleic Acids Res.*, 26, 2016.

61. Bhatia, R. and Verfaillie, C. M. 1998. Inhibition of BCR-ABL expression with antisense oligodeoxynucleotides restores beta1 integrin-mediated adhesion and proliferation inhibition in chronic myelogenous leukemia hematopoietic progenitors, *Blood*, 91, 3414.

62. Bergan, R., Connell, Y., Fahmy, B., and Neckers, L. 1993. Electroporation enhances c-myc antisense oligodeoxynucleotide efficacy, *Nucleic Acids Res.*, 21, 3567.

63. Schiedlmeier, B., Schmitt, R., Muller, W., Kirk, M. M., Gruber, H., Mages, W., and Kirk, D. L. 1994. Nuclear transformation of Volvox carteri, *Proc. Natl. Acad. Sci. USA*, 91, 5080.

64. Lesh, R. E., Somlyo, A. P., Owens, G. K., and Somlyo, A. V. 1995. Reversible permeabilization. A novel technique for the intracellular introduction of antisense oligodeoxynucleotides into intact smooth muscle, *Circ Res*, 77, 220.

65. Spiller, D. G. and Tidd, D. M. 1995. Nuclear delivery of antisense oligodeoxynucleotides through reversible permeabilization of human leukemia cells with streptolysin O, *Antisense Res. Dev.*, 5, 13.

66. Leonetti, J. P., Machy, P., Degols, G., Lebleu, B., and Leserman, L. 1990. Antibody-targeted liposomes containing oligodeoxyribonucleotides complementary to viral RNA selectively inhibit viral replication, *Proc. Natl. Acad. Sci. USA*, 87, 2448.

67. Stein, C. A. and Cheng, Y. C. 1993. Antisense oligonucleotides as therapeutic agents — Is the bullet really magical?, *Science*, 261, 1004.

68. Lisziewicz, J., Sun, D., Klotman, M., Agrawal, S., Zamecnik, P., and Gallo, R. 1992. Specific inhibition of human immunodeficiency virus type 1 replication by antisense oligonucleotides: An *in vitro* model for treatment, *Proc. Natl. Acad. Sci. USA*, 89, 11209.

69. Monia, B. P., Johnston, J. F., Ecker, D. J., Zounes, M. A., Lima, W. F., and Freier, S. M. 1992. Selective inhibition of mutant Ha-ras mRNA expression by antisense oligonucleotides, *J. Biol. Chem.*, 267, 19954.

70. Carter, G. and Lemoine, N. R. 1993. Antisense technology for cancer therapy: Does it make sense?, *Br. J. Cancer*, 67, 869.

71. Chang, E. H., Miller, P. S., Cushman, C., Devadas, K., Pirollo, K. F., Ts'o, P. O., and Yu, Z. P. 1991. Antisense inhibition of *ras* p21 expression that is sensitive to a point mutation, *Biochemistry*, 30, 8283.

72. Agrawal, S., Temsamani, J., and Tang, J. Y. 1991. Pharmacokinetics, biodistribution, and stability of oligodeoxynucleotide phosphorothioates in mice, *Proc. Natl. Acad. Sci. USA*, 88, 7595.

73. Iversen, P. 1991. *In vivo* studies with phosphorothioate oligonucleotides: pharmacokinetics prologue, *Anticancer Drug Des.*, 6, 531.

74. Henry, S. P., Giclas, P. C., Leeds, J., Pangburn, M., Auletta, C., Levin, A. A., and Kornbrust, D. J. 1997. Activation of the alternative pathway of complement by a phosphorothioate oligonucleotide: potential mechanism of action, *J. Pharmacol. Exp. Ther.*, 281, 810.

75. Yu, D., Iyer, R. P., Shaw, D. R., Lisziewicz, J., Li, Y., Jiang, Z., Roskey, A., and Agrawal, S. 1996. Hybrid oligonucleotides: synthesis, biophysical properties, stability studies, and biological activity, *Bioorg. Med. Chem.*, 4, 1685.

76. Wu, C. H., Wilson, J. M., and Wu, G. Y. 1989. Targeting genes: Delivery and persistent expression of a foreign gene driven by mammalian regulatory elements in vivo, *J. Biol. Chem.*, 264, 16985.

77. Ratajczak, M. Z., Kant, J. A., Luger, S. M., Hijiya, N., Zhang, J., Zon, G., and Gewirtz, A. M. 1992. *In vivo* treatment of human leukemia in a scid mouse model with c-myb antisense oligodeoxynucleotides, *Proc. Natl. Acad. Sci. USA*, 89, 11823.

78. Higgins, K. A., Perez, J. R., Coleman, T. A., Dorshkind, K., McComas, W. A., Sarmiento, U. M., Rosen, C. A., and Narayanan, R. 1993. Antisense inhibition of the p65 subunit of NF-kappa B blocks tumorigenicity and causes tumor regression, *Proc. Natl. Acad. Sci. USA*, 90, 9901.

79. Offensperger, W. B., Offensperger, S., Walter, E., Teubner, K., Igloi, G., Blum, H. E., and Gerok, W. 1993. *In vivo* inhibition of duck hepatitis B virus replication and gene expression by phosphorothioate modified antisense oligodeoxynucleotides, *Embo. J.*, 12, 1257.

80. Simons, M., Edelman, E. R., DeKeyser, J. L., Langer, R., and Rosenberg, R. D. 1992. Antisense *c-myb* oligonucleotides inhibit intimal arterial smooth muscle cell accumulation in vivo, *Nature*, 359, 67.

81. Abe, J., Zhou, W., Taguchi, J., Takuwa, N., Miki, K., Okazaki, H., Kurokawa, K., Kumada, M., and Takuwa, Y. 1994. Suppression of neointimal smooth muscle cell accumulation *in vivo* by antisense cdc2 and cdk2 oligonucleotides in rat carotid artery, *Biochem. Biophys. Res. Commun.*, 198, 16.

82. Bennett, M. R., Anglin, S., McEwan, J. R., Jagoe, R., Newby, A. C., and Evan, G. I. 1994. Inhibition of vascular smooth muscle cell proliferation *in vitro* and *in vivo* by *c-myc* antisense oligodeoxynucleotides, *J. Clin. Invest.*, 93, 820.

83. Bennett, C. F., Kornbrust, D., Henry, S., Stecker, K., Howard, R., Cooper, S., Dutson, S., Hall, W., and Jacoby, H. I. 1997. An ICAM-1 antisense oligonucleotide prevents and reverses dextran sulfate sodium-induced colitis in mice, *J. Pharmacol. Exp. Ther.*, 280, 988.

84. Azad, R. F., Brown-Driver, V., Buckheit, R. W., Jr., and Anderson, K. P. 1995. Antiviral activity of a phosphorothioate oligonucleotide complementary to human cytomegalovirus RNA when used in combination with antiviral nucleoside analogs, *Antiviral Res.*, 28, 101.

85. Anderson, K. P., Fox, M. C., Brown-Driver, V., Martin, M. J., and Azad, R. F. 1996. Inhibition of human cytomegalovirus immediate-early gene expression by an antisense oligonucleotide complementary to immediate-early RNA, *Antimicrob. Agents Chemother*, 40, 2004.

Further Information

The book *Antisense Nucleic Acids and Proteins: Fundamentals and Applications*, edited by Joseph N. M. Mol and Alexander R. van der Krol, is a collection of reviews in the use of antisense nucleic acids to modulate or downregulate gene expression. The book *Antisense RNA and DNA*, edited by James A. H. Murray, explores the use of antisense and catalytic nucleic acids for regulating gene expression.

The journal *Antisense and Nucleic Acid Drug Development*, published by Mary Ann Liebert, Inc., presents original research on antisense technology. For subscription information, contact Antisense and Nucleic Acid Drug Development, Mary Ann Liebert, Inc., 2 Madison Avenue, Larchmont, NY 10538; (914) 834-3100, Fax (914) 834-3688, email: info@liebertpub.com. The biweekly journal *Nucleic Acids Research* publishes papers on physical, chemical, and biologic aspects of nucleic acids, their constituents, and proteins with which they interact. For subscription information, contact Journals Marketing; Oxford University Press, Inc., 2001 Evans Road, Cary, NC 27513; 1-800-852-7323, Fax, (919) 677-1714, email: jnlorders@oup-usa.org.

11

Tools for
Genome Analysis

Robert Kaiser

University of Washington

The development of sophisticated and powerful recombinant techniques for manipulating and analyzing genetic material has led to the emergence of a new biologic discipline, often termed *molecular biology*. The tools of molecular biology have enabled scientists to begin to understand many of the fundamental processes of life, generally through the identification, isolation, and structural and functional analysis of individual or, at best, limited numbers of genes. Biology is now at a point where it is feasible to begin a more ambitious endeavor — the complete genetic analysis of entire genomes. Genome analysis aims not only to identify and molecularly characterize all the genes that orchestrate the development of an organism but also to understand the complex and interactive regulatory mechanisms of these genes, their organization in the genome, and the role of genetic variation in disease, adaptability, and individuality. Additionally, the study of homologous genetic regions across species can provide important insight into their evolutionary history.

As can be seen in Table 11.1, the genome of even a small organism consists of a very large amount of information. Thus the analysis of a complete genome is not simply a matter of using conventional techniques that work well with individual genes (comprised of perhaps 1000 to 10,000 base pairs) a sufficient (very large) number of times to cover the genome. Such a brute-force approach would be too slow and too expensive, and conventional data-handling techniques would be inadequate for the task of cataloging, storing, retrieving, and analyzing such a large amount of information. The amount of manual labor and scientific expertise required would be prohibitive. New technology is needed to provide high-throughput, low-cost automation, and reduced reliance on expert intervention at intermediate levels in the processes required for large-scale genetic analysis. Novel computational tools are required to deal with the large volumes of genetic information produced. Individual tools must be integrated smoothly to produce an analytical system in which samples are tracked through the entire analytical process, intermediate decisions and branch points are few, a stable, reliable, and routine protocol or set of protocols is employed, and the resulting information is presented to the biologic scientist in a useful and meaningful format. It is important to realize that the development of these tools requires the interdisciplinary efforts of biologists, chemists, physicists, engineers, mathematicians, and computer scientists.

Genome analysis is a complex and extended series of interrelated processes. The basic processes involved are diagrammed in Fig. 11.1. At each stage, new biologic, chemical, physical (mechanical, optical), and computational tools have been developed within the last 10 years that have begun to enable large-scale (megabase) genetic analysis. These developments have largely been spurred by the goals of the Human Genome Project, a worldwide effort to decipher the entirety of human genetics. However,

0-8493-1811-4/03/$0.00+$1.50

TABLE 11.1   DNA Content of Various Genomes in Monomer Units (Base Pairs)

Organism	Type	Size
Phage T4	Bacteriophage (virus)	160,000
Escherichia coli	Bacterium	4,000,000
Saccharomyces	Yeast	14,000,000
Arabidopsis thaliana	Plant	100,000,000
Caenorhabditis elegans	Nematode	100,000,000
Drosophila melanogaster	Insect (fruit fly)	165,000,000
Mouse	Mammal	3,000,000,000
Human	Mammal	3,500,000,000

Source: Adapted from [1] and [2].

biologists are still a significant distance away from having a true genome analysis capability [3], and as such, new technologies are still emerging.

This chapter cannot hope to describe in depth the entire suite of tools currently in use in genome analysis. Instead, it will attempt to present the basic principles involved and to highlight some of the recent enabling technological developments that are likely to remain in use in genome analysis for the foreseeable future. Some fundamental knowledge of biology is assumed; in this regard, an excellent beginning text for individuals with a minimal background in molecular biology is that by Watson et al. [1].

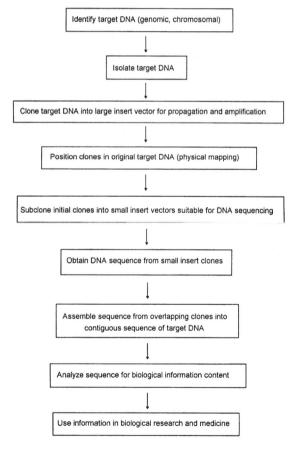

FIGURE 11.1   Basic steps in genome analysis.

TABLE 11.2 Enzymes Commonly Used in Genome Analysis

Enzyme	Function	Common Use
Restriction endonuclease	Cleave double-stranded DNA at specific sites	Mapping, cloning
DNA polymerase	Synthesize complementary DNA strand	DNA sequencing, amplification
Polynucleotide kinase	Adds phosphate to 5′ end of single-stranded DNA	Radiolabeling, cloning
Terminal transferase	Adds nucleotides to the 3′ end of single-stranded DNA	Labeling
Reverse transcriptase	Makes DNA copy from RNA	RNA sequencing, cDNA cloning
DNA ligase	Covalently joins two DNA fragments	Cloning

11.1 General Principles

The fundamental blueprint for any cell or organism is encoded in its genetic material, its deoxyribonucleic acid (DNA). DNA is a linear polymer derived from a four-letter biochemical alphabet — A, C, G, and T. These four letters are often referred to as *nucleotides* or *bases*. The linear order of bases in a segment of DNA is termed its *DNA sequence* and determines its function. A gene is a segment of DNA whose sequence directly determines its translated protein product. Other DNA sequences are recognized by the cellular machinery as start-and-stop sites for protein synthesis, regulate the temporal or spatial expression of genes, or play a role in the organization of higher-order DNA structures such as chromosomes. Thus a thorough understanding of the DNA sequence of a cell or organism is fundamental to an understanding of its biology.

Recombinant DNA technology affords biologists the capability to manipulate and analyze DNA sequences. Many of the techniques employed take advantage of a basic property of DNA, the molecular complementarity of the two strands of the double helix. This complementarity arises from the specific hydrogen-bonding interactions between pairs of DNA bases, A with T and C with G. Paired double strands of DNA can be denatured, or rendered into the component single strands, by any process that disrupts these hydrogens bonds — high temperature, chaotropic agents, or pH extremes. Complementary single strands also can be renatured into the duplex structure by reversing the disruptive element; this process is sometimes referred to as *hybridization* or *annealing*, particularly when one of the strands has been supplied from some exogenous source.

Molecular biology makes extensive use of the DNA-modifying enzymes employed by cells during replication, translation, repair, and protection from foreign DNA. A list of commonly used enzymes, their functions, and some of the experimental techniques in which they are utilized is provided in Table 11.2.

11.2 Enabling Technologies

The following are broadly applicable tools that have been developed in the context of molecular biology and are commonly used in genome analysis.

Cloning. Cloning is a recombinant procedure that has two main purposes: First, it allows selection of a single DNA fragment from a complex mixture, and second, it provides a means to store, manipulate, propagate, and produce large numbers of identical molecules having this single ancestor. A cloning vector is a DNA fragment derived from a microorganism, such as a bacteriophage or yeast, into which a foreign DNA fragment may be inserted to produce a chimeric DNA species. The vector contains all the genetic information necessary to allow for the replication of the chimera in an appropriate host organism. A variety of cloning vectors have been developed that allow for the insertion and stable propagation of foreign DNA segments of various sizes; these are indicated in Table 11.3.

Electrophoresis. Electrophoresis is a process whereby nucleic acids are separated by size in a sieving matrix under the influence of an electric field. In free solution, DNA, being highly negatively charged by virtue of its phosphodiester backbone, migrates rapidly toward the positive pole of an electric field. If the DNA is forced instead to travel through a molecularly porous substance, such as a gel, the smaller

TABLE 11.3 Common Cloning Vectors

Vector	Approximate Insert Size Range (Base Pairs)
Bacteriophage M13	100–5000
Plasmid	100–10,000
Bacteriophage lambda	10,000–15,000
Cosmid	25,000–50,000
Yeast artificial chromosome (YAC)	100,000–1,000,000

(shorter) fragments of DNA will travel through the pores more rapidly than the larger (longer) fragments, thus effecting separation. Agarose, a highly purified derivative of agar, is commonly used to separate relatively large fragments of DNA (100 to 50,000 base pairs) with modest resolution (50 to 100 base pairs), while cross-linked polyacrylamide is used to separate smaller fragments (10 to 1,000 base pairs) with single base-pair resolution. Fragment sizes are generally estimated by comparison with standards run in another lane of the same gel. Electrophoresis is used extensively as both an analytical and a preparative tool in molecular biology.

Enzymatic DNA Sequencing.   In the late 1970s, Sanger and coworkers [4] reported a procedure employing DNA polymerase to obtain DNA sequence information from unknown cloned fragments. While significant improvements and modifications have been made since that time, the basic technique remains the same: DNA polymerase is used to synthesize a complementary copy of an unknown single-stranded DNA (the template) in the presence of the four DNA monomers (deoxynucleotide triphosphates, or dNTPs). DNA polymerase requires a double-stranded starting point, so a single-stranded DNA (the primer) is hybridized at a unique site on the template (usually in the vector), and it is at this point that DNA synthesis is initiated. Key to the sequencing process is the use of a modified monomer, a dideoxy-nucleotide triphosphate (ddNTP), in each reaction. The ddNTP lacks the 3'-hydroxyl functionality (it has been replaced by a hydrogen) necessary for phosphodiester bond formation, and its incorporation thus blocks further elongation of the growing chain by polymerase. Four reactions are carried out, each containing all four dNTPs and one of the four ddNTPs. By using the proper ratios of dNTPs to ddNTP, each reaction generates a nested set of fragments, each fragment beginning at exactly the same point (the primer) and terminating with a particular ddNTP at each base complementary to that ddNTP in the template sequence. The products of the reactions are then separated by electrophoresis in four lanes of a polyacrylamide slab gel. Since conventional sequencing procedures utilize radiolabeling (incorporation of a small amount of ^{32}P- or ^{35}S-labeled dNTP by the polymerase), visualization of the gel is achieved by exposing it to film. The sequence can be obtained from the resulting autoradiogram, which appears as a series of bands (often termed a *ladder*) in each of the four lanes. Each band is composed of fragments of a single size, the shortest fragments being at the bottom of the gel and the longest at the top. Adjacent bands represent a single base pair difference, so the sequence is determined by reading up the ladders in the four lanes and noting which lane contains the band with the next largest sized fragments. The enzymatic sequencing process is diagrammed in Fig. 11.2. It should be noted that although other methods exist, the enzymatic sequencing technique is currently the most commonly used DNA sequencing procedure due to its simplicity and reliability.

Polymerase Chain Reaction (PCR).   PCR [5] is an *in vitro* procedure for amplifying particular DNA sequences up to 10^8-fold that is utilized in an ever-increasing variety of ways in genome analysis. The sequence to be amplified is defined by a pair of single-stranded primers designed to hybridize to unique sites flanking the target sequence on opposite strands. DNA polymerase in the presence of the four dNTPs is used to synthesize a complementary DNA copy across the target sequence starting at the two primer sites. The amplification procedure is performed by repeating the following cycle 25 to 50 times (see Fig. 11.3). First, the double-stranded target DNA is denatured at high temperature (94 to 96°C). Second, the mixture is cooled, allowing the primers to anneal to their complementary sites on the target single strands. Third, the temperature is adjusted for optimal DNA polymerase activity, initiating synthesis. Since the primers are complementary to the newly synthesized strands as well as the original target, each

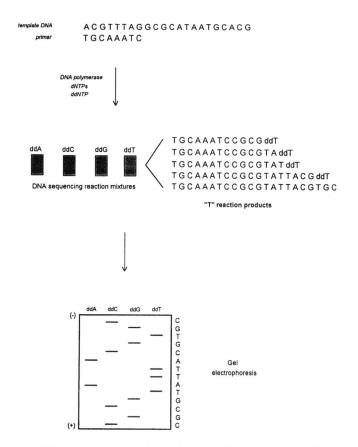

FIGURE 11.2 Enzymatic DNA sequencing. A synthetic oligonucleotide primer is hybridized to its complementary site on the template DNA. DNA polymerase and dNTPs are then used to synthesize a complementary copy of the unknown portion of the template in the presence of a chain-terminating ddNTP (see text). A nested set of fragments beginning with the primer sequence and ending at every ddNTP position is produced in each reaction (the ddTTP reaction products are shown). Four reactions are carried out, one for each ddNTP. The products of each reaction are then separated by gel electrophoresis in individual lanes, and the resulting ladders are visualized. The DNA sequence is obtained by reading up the set of four ladders, one base at a time, from smallest to largest fragment.

cycle of denaturation/annealing/synthesis effectively doubles the amount of target sequence present in the reaction, resulting in a 2^n amplification (n = number of cycles). The initial implementation of PCR utilized a polymerase that was unstable at the high temperatures required for denaturation, thus requiring manual addition of polymerase prior to the synthesis step of every cycle. An important technological development was the isolation of DNA polymerase from a thermophilic bacterium, *Thermus aquaticus* (*Taq*), which can withstand the high denaturation temperatures [6]. Additionally, the high optimal synthesis temperature (70 to 72°C) of *Taq* polymerase improves the specificity of the amplification process by reducing spurious priming from annealing of the primers to nonspecific secondary sites in the target.

While PCR can be performed successfully manually, it is a tedious process, and numerous thermal cycling instruments have become commercially available. Modern thermal cyclers are programmable and capable of processing many samples at once, using either small plastic tubes or microtiter plates, and are characterized by accurate and consistent temperature control at all sample positions, rapid temperature ramping, and minimal temperature over/undershoot. Temperature control is provided by a variety of means (Peltier elements, forced air, water) using metal blocks or water or air baths. Speed, precise temperature control, and high sample throughput are the watchwords of current thermal cycler design.

PCR technology is commonly used to provide sufficient material for cloning from genomic DNA sources, to identify and characterize particular DNA sequences in an unknown mixture, to rapidly

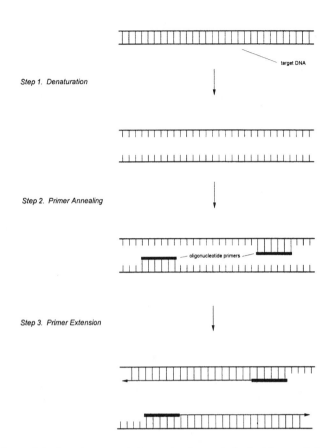

Step 1. Denaturation

Step 2. Primer Annealing

Step 3. Primer Extension

FIGURE 11.3 The first cycle in the polymerase chain reaction. In step 1, the double-stranded target DNA is thermally denatured to produce single-stranded species. A pair of synthetic primers, flanking the specific region of interest, are annealed to the single strands to form initiation sites for DNA synthesis by polymerase (step 2). Finally, complementary copies of each target single strand are synthesized by polymerase in the presence of dNTPs, thus doubling the amount of target DNA initially present (step 3). Repetition of this cycle effectively doubles the target population, affording one million-fold or greater amplification of the initial target sequence.

produce templates for DNA sequencing from very small amounts of target DNA, and in cycle sequencing, a modification of the enzymatic sequencing procedure that utilizes *Taq* polymerase and thermal cycling to amplify the products of the sequencing reactions.

Chemical Synthesis of Oligodeoxynucleotides.   The widespread use of techniques based on DNA polymerase, such as enzymatic DNA sequencing and the PCR, as well as of numerous other techniques utilizing short, defined-sequence, single-stranded DNAs in genome analysis, is largely due to the ease with which small oligodeoxynucleotides can be obtained. The chemical synthesis of oligonucleotides has become a routine feature of both individual biology laboratories and core facilities. The most widely used chemistry for assembling short (<100 base pair) oligonucleotides is the phosphoramidite approach [7], which has developed over the past 15 years or so. This approach is characterized by rapid, high-yield reactions and stable reagents. Like modern peptide synthesis chemistry, the approach relies on the tethering of the growing DNA chain to a solid support (classically glass or silica beads, more recently cross-linked polystyrene) and is cyclic in nature. At the end of the assembly, the desired oligonucleotide is chemically cleaved from the support, generally in a form that is sufficiently pure for its immediate use in a number of applications. The solid phase provides two significant advantages: It allows for the use of large reagent excesses, driving the reactions to near completion in accord with the laws of mass action while reducing the removal of these excesses following the reactions to a simple matter of thorough washing, and it enables the reactions to be performed in simple, flow-through cartridges, making the entire synthesis procedure easily automatable. Indeed, a number of chemical DNA synthesis instruments ("gene machines") are commercially available, capable of synthesizing one to several oligonucleotides at once. Desired sequences are programmed through a keyboard or touchpad, reagents are installed, and DNA is obtained a few hours later. Improvements in both chemistry and instrument design have been aimed at increasing synthesis throughput (reduced cycle times, increased number of simultaneous sequence assemblies), decreasing scale (most applications in genome analysis require subnanomole quantities of any particular oligonucleotide), and concomitant with these two, reducing cost per oligonucleotide.

11.3 Tools for Genome Analysis

Physical Mapping. In the analysis of genomes, it is often useful to begin with a less complex mixture than an entire genome DNA sample. Individual chromosomes can be obtained in high purity using a technology known as *chromosome sorting* [8], a form of flow cytometry. A suspension of chromosomes stained with fluorescent dye is flowed past a laser beam. As a chromosome enters the beam, appropriate optics detect the scattered and emitted light. Past the beam, the stream is acoustically broken into small droplets. The optical signals are used to electronically trigger the collection of droplets containing chromosomes by electrostatically charging these droplets and deflecting them into a collection medium using a strong electric field. Chromosomes can be differentiated by staining the suspension with two different dyes that bind in differing amounts to the various chromosomes and looking at the ratio of the emission intensity of each dye as it passes the laser/detector. Current commercial chromosome sorting instrumentation is relatively slow, requiring several days to collect sufficient material for subsequent analysis.

As mentioned previously, whole genomes or even chromosomes cannot yet be analyzed as intact entities. As such, fractionation of large nucleic acids into smaller fragments is necessary to obtain the physical material on which to perform genetic analysis. Fractionation can be achieved using a variety of techniques: limited or complete digestion by restriction enzymes, sonication, or physical shearing through a small orifice. These fragments are then cloned into an appropriate vector, the choice of which depends on the size range of fragments involved (see Table 11.3). The composite set of clones derived from a large nucleic acid is termed a *library*. In general, it is necessary to produce several libraries in different cloning vectors containing different-sized inserts. This is necessary because the mapping of clones is facilitated by larger inserts, while the sequencing of clones requires shorter inserts.

The library-generating process yields a very large number of clones having an almost random distribution of insert endpoints in the original fragment. It would be very costly to analyze all clones in a library, and unnecessary as well. Instead, a subset of overlapping clones is selected whose inserts span the entire starting fragment. These clones must be arrayed in the linear order in which they are found in the starting fragment; the process for doing this is called *physical mapping*. The conventional method for physically mapping clones uses restriction enzymes to cleave each clone at enzyme-specific sites, separating the products of the digestion by electrophoresis, and comparing the resulting patterns of restriction fragment sizes for different clones to find similarities. Clones exhibiting a number of the same-sized fragments likely possess the same subsequence and thus overlap. Clearly, the longer the inserts contained in the library, the faster a large genetic region can be covered by this process, since fewer clones are required to span the distance. Physical mapping also provides landmarks, the enzyme cleavage sites in the sequence, that can be used to provide reference points for the mapping of genes and other functional sequences. Mapping by restriction enzyme digestion is simple and reliable to perform; however, manual map assembly from the digest data is laborious, and significant effort is currently being expended in the development of robust and accurate map assembly software.

Normal agarose gel electrophoresis can effectively separate DNA fragments less than 10,000 base pairs and fragments between 10,000 and 50,000 base pairs less effectively under special conditions. However, the development of very large insert cloning vectors, such as the yeast artificial chromosome [9], necessitated the separation of fragments significantly larger than 10,000 base pairs to allow for use in physical mapping. In order to address this issue, a technology called *pulsed-field gel electrophoresis*

(PFGE) was developed. Unlike conventional electrophoresis, in which the electric field remains essentially constant, homogeneous, and unidirectional during a separation, PFGE utilizes an electric field that periodically changes its orientation. The principle of PFGE is thought to be as follows: When DNA molecules are placed in an electric field, the molecules elongate in the direction of the field and then begin to migrate through the gel pores. When the field is removed, the molecules relax to a more random coiled state and stop moving. Reapplication of the field in another orientation causes the DNA to change its conformation in order to align in that direction prior to migration. The time required for this conformational change to occur has been found to be very dependent on the size of the molecules, with larger molecules reorienting more slowly than small ones. Thus longer DNAs move more slowly under the influence of the constantly switching electric field than shorter ones, and size-based separation occurs. PFGE separations of molecules as large as 10 million base pairs have been demonstrated. Numerous instruments for PFGE have been constructed, differing largely in the strategy employed to provide electric field switching [10].

Physical maps based on restriction sites are of limited long-term utility, since they require the provision of physical material from the specific library from which the map was derived in order to be utilized experimentally. A more robust landmarking approach based on the PCR has been developed recently [11], termed *sequence-tagged site* (STS) *content mapping*. An STS is a short, unique sequence in a genome that can be amplified by the PCR. Clones in a library are screened for the presence of a particular STS using PCR; if the STS is indeed unique in the genome, then clones possessing that STS are reliably expected to overlap. Physical mapping is thus reduced to choosing and synthesizing pairs of PCR primers that define unique sequences in the genome. Additionally, since STSs are defined by pairs of primer sequences, they can be stored in a database and are thus universally accessible.

DNA Sequencing.   Early in the development of tools for large-scale DNA analysis, it was recognized that one of the most costly and time-consuming processes was the accumulation of DNA sequence information. Two factors, the use of radioisotopic labels and the manual reading and recording of DNA sequence films, made it impossible to consider genome-scale (10^6 to 10^9 base pairs) sequence analysis using the conventional techniques. To address this, several groups embarked on the development of automated DNA sequencing instruments [12–14]. Today, automated DNA sequencing is one of the most highly advanced of the technologies for genome analysis, largely due to the extensive effort expended in instrument design, biochemical organization, and software development.

Key to the development of these instruments was the demonstration that fluorescence could be employed in the place of autoradiography for detection of fragments in DNA sequencing gels and that the use of fluorescent labels enabled the acquisition of DNA sequence data in an automated fashion in real time. Two approaches have been demonstrated: the "single-color, four-lane" approach and the "four-color, single-lane" approach. The former simply replaces the radioisotopic label used in conventional enzymatic sequencing with a fluorescent label, and the sequence is determined by the order of fluorescent bands in the four lanes of the gel. The latter utilizes a different-colored label for each of the four sequencing reactions (thus A might be "blue," C, "green," G, "yellow," and T, "red"). The four base-specific reactions are performed separately and upon completion are combined and electrophoresed in a single lane of the gel, and the DNA sequence is determined from the temporal pattern of fluorescent colors passing the detector. For a fixed number of gel lanes (current commercial automated DNA sequencers have 24 to 36), the four-color approach provides greater sample throughput than the single-color approach. Instruments employing the four-color technology are more widely used for genome analysis at present, and as such, this strategy will be discussed more fully.

In order to utilize fluorescence as a detection strategy for DNA sequencing, a chemistry had to be developed for the specific incorporation of fluorophores into the nested set of fragments produced in the enzymatic sequencing reactions. The flexibility of chemical DNA synthesis provided a solution to this problem. A chemistry was developed for the incorporation of an aliphatic primary amine in the last cycle of primer synthesis (i.e., at the 5′ terminus) using standard DNA synthesis protocols [15,16]. This amine was then conjugated with any of several of readily available amine-reactive fluorochromes that had been developed previously for the labeling of proteins to produce the desired labeled sequencing

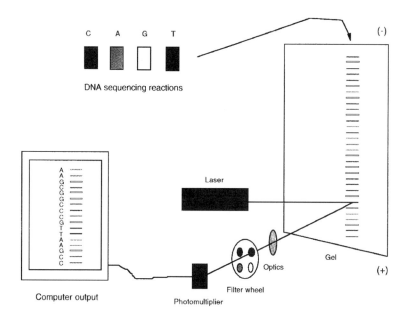

FIGURE 11.4 Schematic illustration of automated fluorescence-based DNA sequencing using the "four-color, single-lane" approach. The products of each of the four enzymatic sequencing reactions are "color-coded" with a different fluorescent dye, either through the use of dye-labeled primers or dye-terminators. The four reaction mixtures are then combined and the mixture separated by gel electrophoresis. The beam of an argon ion laser is mechanically scanned across the width of the gel near its bottom to excite the labeled fragments to fluorescence. The emitted light is collected through a four-color filter wheel onto a photomultiplier tube. The color of each fluorescing band is determined automatically by a computer from the characteristic four-point spectrum of each dye, and the order of colors passing the detector is subsequently translated into DNA sequence information.

primers. The purified dye-primer was demonstrated to perform well in DNA sequencing, exhibiting both efficient extension by the polymerase and the necessary single-base resolution in the electrophoretic separation [12]. A set of four spectrally discriminable, amine-reactive fluorophores has been developed [16] for DNA sequencing.

While dye-primers are relatively easy to obtain by this method, they are costly to prepare in small quantities and require sophisticated chromatographic instrumentation to obtain products pure enough for sequencing use. Thus they are generally prepared in large amounts and employed as vector-specific "universal" primers, as in situations in which a very large number of inserts cloned in a given vector need to be sequenced [17]. For occasions where small amounts of sample-specific primers are needed, as in the sequencing of products from the PCR, a simpler and more economical alternative is the use of dideoxynucleotides covalently coupled to fluorescent dyes (so-called dye-terminators), since these reagents allow the use of conventional unlabeled primers [18].

Special instrumentation (Fig. 11.4) has been developed for the fluorescence-based detection of nucleic acids in DNA sequencing gels. An argon ion laser is used to excite the fluorescent labels in order to provide sufficient excitation energy at the appropriate wavelength for high-sensitivity detection. The laser beam is mechanically scanned across the width of the gel near its bottom in order to interrogate all lanes. As the labeled DNA fragments undergoing electrophoresis move through the laser beam, their emission is collected by focusing optics onto a photomultiplier tube located on the scanning stage. Between the photomultiplier tube and the gel is a rotating four-color filter wheel. The emitted light from the gel is collected through each of the four filters in the wheel in turn, generating a continuous four-point spectrum of the detected radiation. The color of the emission of the passing bands is determined from the characteristic four-point spectrum of each fluorophore, and the identified color is then translated into DNA sequence using the associated dye/base pairings. Sequence acquisition and data analysis are handled completely by computer; the system is sufficiently sophisticated that the operator can load the samples on the gel, activate the electrophoresis and the data acquisition, and return the next day for the analyzed data. The current commercial implementation of this technology produces about 450 to 500 bases per lane of analyzed DNA sequence information at an error rate of a few percent in a 12- to 14-hour period.

The rate of data production of current DNA sequencers is still too low to provide true genome-scale analytical capabilities, although projects in the few hundred kilobase pair range have been accomplished. Improvements such as the use of gel-filled capillaries [19] or ultrathin slab gels (thicknesses on the order of 50 to 100 μm as opposed to the conventional 200 to 400 μm) [20,21] are currently being explored. The improved heat dissipation in these thin-gel systems allows for the use of increased electric field strengths during electrophoresis, with a concomitant reduction in run time of fivefold or so.

The greatly increased throughput of the automated instruments over manual techniques has resulted in the generation of a new bottleneck in the DNA sequencing process that will only be exacerbated by higher-throughput systems: the preparation of sufficient sequencing reaction products for analysis. This encompasses two preparative processes, the preparation of sequencing templates and the performance of sequencing reactions. Automation of the latter process has been approached initially through the development of programmable pipetting robots that operate in the 96-well microtiter plate format commonly used for immunoassays, since a 96-well plate will accommodate sequencing reactions for 24 templates. *Sequencing robots* of this sort have become important tools for large-scale sequencing projects. Template preparation has proven more difficult to automate. No reliable system for selecting clones, infecting and culturing bacteria, and isolating DNA has been produced, although several attempts are in progress. It is clear that unlike in the case of the sequencing robots, where instrumentation has mimicked manual manipulations with programmable mechanics, successful automation of template preparation will require a rethinking of current techniques with an eye toward process automation. Furthermore, in order to obtain true genome-scale automation, the entire sequencing procedure from template preparation through data acquisition will need to be reengineered to minimize, if not eliminate, operator intervention using the principles of systems integration, process management, and feedback control.

Genetic Mapping. Simply stated, genetic mapping is concerned with identifying the location of genes on chromosomes. Classically, this is accomplished using a combination of mendelian and molecular genetics called *linkage analysis*, a complete description of which is too complex to be fully described here (but see Watson et al. [1]). However, an interesting approach to physically locating clones (and the genes contained within them) on chromosomes is afforded by a technique termed *fluorescence in situ hybridization* (FISH) [22]. Fluorescently labeled DNA probes derived from cosmid clones can be hybridized to chromosome spreads, and the location of the probe-chromosome hybrid can be observed using fluorescence microscopy. Not only can clones be mapped to particular chromosomes in this way, but positions and distances relative to chromosomal landmarks (such as cytogenetic bands, telomere, or centromere) can be estimated to as little as 50,000 base pairs in some cases, although 1 million base pairs or larger is more usual. The technique is particularly useful when two or more probes of different colors are used to order sequences relative to one another.

Computation. Computation plays a central role in genome analysis at a variety of levels, and significant efforts has been expanded on the development of software and hardware tools for biologic applications. A large effort has been expended in the development of software that will rapidly assemble a large contiguous DNA sequence from the many smaller sequences obtained from automated instruments. This assembly process is computationally demanding, and only recently have good software tools for this purpose become readily available. Automated sequencers produce 400 to 500 base pairs of sequence per template per run. However, in order to completely determine the linear sequence of a 50,000-base-pair cosmid insert (which can be conceptually represented as a linear array of 100 adjacent 500-base-pair templates), it is necessary to assemble sequence from some 300 to 1000 clones to obtain the redundancy of data needed for a high-accuracy finished sequence, depending on the degree to which the clones can be preselected for sequencing based on a previously determined physical map. Currently, the tools for acquiring and assembling sequence information are significantly better than those for physical mapping; as such, most large-scale projects employ strategies that emphasize sequencing at the expense of mapping [23]. Improved tools for acquiring and assembling mapping data are under development, however, and it remains to be seen what effect they will have on the speed and cost of obtaining finished sequence on a genome scale relative to the current situation.

Many software tools have been developed in the context of the local needs of large-scale projects. These include software for instrument control (data acquisition and signal processing), laboratory information-management systems, and local data-handling schemes. The development of process approaches to the automation of genome analysis will necessitate the continued development of tools of these types.

The final outcome of the analysis of any genome will be a tremendous amount of sequence information that must be accessible to researchers interested in the biology of the organism from which it was derived. Frequently, the finished genomic sequence will be the aggregate result of the efforts of many laboratories. National and international information resources (databases) are currently being established worldwide to address the issues of collecting, storing, correlating, annotating, standardizing, and distributing this information. Significant effort is also being expended to develop tools for the rapid analysis of genome sequence data that will enable biologists to find new genes and other functional genetic regions, compare very large DNA sequences for similarity, and study the role of genetic variation in biology. Eventually, as the robust tools for predicting protein tertiary structure and function from primary acid sequence data are developed, genome analysis will extend to the protein domain through the translation of new DNA sequences into their functional protein products.

11.4 Conclusions

Genome analysis is a large-scale endeavor whose goal is the complete understanding of the basic blueprint of life The scale of even the smallest genomes of biologic interest is too large to be effectively analyzed using the traditional tools of molecular biology and genetics. Over the past 10 years, a suite of biochemical techniques and bioanalytical instrumentation has been developed that has allowed biologists to begin to probe large genetic regions, although true genome-scale technology is still in its infancy. It is anticipated that the next 10 years will see developments in the technology for physical mapping, DNA sequencing, and genetic mapping that will allow for a 10- to 100-fold increase in our ability to analyze genomes, with a concomitant decrease in cost, through the application of process-based principles and assembly-line approaches. The successful realization of a true genome analysis capability will require the close collaborative efforts of individuals from numerous disciplines in both science and engineering.

Acknowledgments

I would like to thank Dr. Leroy Hood, Dr. Maynard Olson, Dr. Barbara Trask, Dr. Tim Hunkapiller, Dr. Deborah Nickerson, and Dr. Lee Rowen for the useful information, both written and verbal, that they provided during the preparation of this chapter.

References

1. Watson JD, Gilman M, Witkowski J, Zoller M (eds). 1992. *Recombinant DNA*, 2d ed. New York, Scientific American Books, WH Freeman.
2. Lewin B. 1987. *Genes III*, 3d ed. New York, Wiley.
3. Olson MV. 1993. The human genome project. *Proc Natl Acad Sci USA* 90:4338.
4. Sanger F, Nicklen S, Coulson AR. 1977. DNA sequencing with chain-terminating inhibitors. *Proc Natl Acad Sci USA* 74:5463.
5. Saiki RK, Scharf SJ, Faloona F, et al. 1985. Enzymatic amplification of betaglobin sequences and restriction site analysis for diagnosis of sickle cell anemia. *Science* 230:1350.

6. Saiki RK, Gelfand DH, Stoffel S, et al. 1988. Primer-directed enzymatic amplification of DNA with a thermostable DNA polymerase. *Science* 239:487.

7. Gait MJ (ed). 1984. *Oligonucleotide Synthesis: A Practical Approach*. Oxford, England, IRL Press.

8. Engh Gvd. 1993. New applications of flow cytometry. *Curr Opin Biotechnol* 4:63.

9. Burke DT, Carle GF, Olson MV. 1987. Cloning of large segments of exogenous DNA into yeast by means of artificial chromosome vectors. *Science* 236:806.

10. Lai E, Birren BW, Clark SM, Hood L. 1989. Pulsed field gel electrophoresis. *Biotechniques* 7:34.

11. Olson M, Hood L, Cantor C, Botstein D. 1989. A common language for physical mapping of the human genome. *Science* 245:1434.

12. Smith LM, Sanders JZ, Kaiser RJ, et al. 1986. Fluorescence detection in automated DNA sequence analysis. *Nature* 321:674.

13. Prober JM, Trainor GL, Dam RJ, et al. 1987. A system for rapid DNA sequencing with fluorescent chain terminating dideoxynucleotides. *Science* 238:336.

14. Ansorge W, Sproat B, Stegemann J, et al. 1987. Automated DNA sequencing: Ultrasensitive detection of fluorescent bands during electrophoresis. *Nucleic Acid Res* 15:4593.

15. Smith LM, Fung S, Hunkapiller MW, et al. 1985. The synthesis of oligonucleotides containing an aliphtic amino group at the 5′ terminus: Synthesis of fluorescent DNA primers for use in DNA sequencing. *Nucleic Acids Res* 15:2399.

16. Connell C, Fung S, Heiner C, et al. 1987. *Biotechniques* 5:342.

17. Kaiser R, Hunkapiller T, Heiner C, Hood L. 1993. Specific primer-directed DNA sequence analysis using automated fluorescence detection and labeled primers. *Methods Enzymol* 218:122.

18. Lee LG, Connell CR, Woo SL, et al. 1992. DNA sequencing with dye-labeled terminators and T7 DNA polymerase: Effect of dyes and dNTPs on incorporation of dye-terminators and probability analysis of termination fragments. *Nucleic Acids Res* 20:2471.

19. Mathies RA, Huang XC. 1992. Capillary array electrophoresis: an approach to high-speed, high-throughput DNA sequencing. *Nature* 359:167.

20. Brumley RL Jr, Smith LM. 1991. Rapid DNA sequencing by horizontal ultrathin gel electrophoresis. *Nucleic Acid Res* 19:4121.

21. Stegemann J, Schwager C, Erfle H, et al. 1991. High speed on-line DNA sequencing on ultrathin slab gels. *Nucleic Acid Res* 19:675.

22. Trask BJ. 1991. Gene mapping by in situ hybridization. *Curr Opin Genet Dev* 1:82.

23. Hunkapiller T, Kaiser RJ, Koop BF, Hood L. 1991. Large-scale and automated DNA sequencing. *Science* 254:59.

12

Vaccine Production

John G. Aunins
Merck Research Laboratories

Ann L. Lee
Merck Research Laboratories

David B. Volkin
Merck Research Laboratories

Vaccines are biologic preparations that elicit immune system responses that protect an animal against pathogenic organisms. The primary component of the vaccine is an antigen, which can be a weakened (attenuated) version of an infectious pathogen or a purified molecule derived from the pathogen. Upon oral administration or injection of a vaccine, the immune system generates humoral (antibody) and cellular (cytotoxic, or killer T cell) responses that destroy the antigen or antigen-infected cells. When properly administered, the immune response to a vaccine has a long-term memory component, which protects the host against future infections. Vaccines often contain adjuvants to enhance immune response, as well as formulation agents to preserve the antigen during storage or upon administration, to provide proper delivery of antigen, and to minimize side reactions.

Table 12.1 presents a simple classification scheme for vaccines according to the type of organism and antigen, *Live, attenuated whole-organism vaccines* have been favored for simplicity of manufacture and for the strong immune response generated when the organism creates a subclinical infection before being overwhelmed. These are useful when the organism can be reliably attenuated in pathogenicity (while maintaining immunogenicity) or when the organism is difficult to cultivate *ex vivo* in large quantities and hence large amounts of antigen cannot be prepared. Conversely, *subunit antigen vaccines* are used when it is easy to generate large amounts of the antigen or when the whole organisms is not reliably attenuated. Since there is no replication *in vivo*, subunit vaccines rely on administrating relatively large amounts of antigen mass and are almost always adjuvanted to try to minimize the antigen needed. Subunit preparations have steadily gained favor, since biologic, engineering, and analytical improvements make them easily manufactured and characterized to a consistent standard. *Passive vaccines* are antibody preparations from human blood serum. These substitute for the patient's humoral response for immune-suppressed persons, for postexposure prophylaxis of disease, and for high-infection risk situations where immediate protection is required, such as for travelers or medical personnel. Even more so than subunit vaccines, large amounts of antibodies are required. These vaccines do not provide long-term immune memory.

The organism and the nature of the antigen combine to determine the technologies of manufacture for the vaccine. Vaccine production generally involves growing the organism or its antigenic parts (cultivation), treating it to purify and/or detoxify the organism and antigen (downstream processing),

TABLE 12.1   Examples of Vaccines against Human Pathogens, and Their Classification

Type of Organism	Live, Attenuated Cells/Particles	Killed, Inactivated Cells/Particles	Subcellular Particles/Vaccines	Passive Immunization
Virus	Measles, mumps, rubella, polio, yellow fever, varicella,* rotavirus*	Polio, hepatitis B, rabies, yellow fever, Japanese encephalitis, influenza, hepatitis A*	*Virus-like particles* Hepatitis B, HIV gp120-vaccinia* *Characterized single/combined antigens* Influenza HA + NA, herpes simplex gD*	*Immune serum globulin* Rabies, hepatitis B, cytomegalovirus *Monoclonal antibody* Anti-HIV gp120*
Bacterium	Tuberculosis (BCG), typhoid	Pertussis, cholera, plague bacillus	Toxoids Tetanus, diphtheria Characterized single/combined antigens Pertussis PT + FHA + LPF + PSA + 69kD, *Pneumococcus* polysaccharides (23), *Meningococcus* polysaccharides Polysaccharide-protein conjugates *Haemophilus* b, *Pneumococcus**	Immune serum globulin Tetanus, Nonspecific Ig

* Denotes vaccines in development.

and in many cases further combining the antigen with adjuvants to increase its antigenicity and improve storage stability (formulation). These three aspects of production will be discussed, in addition to future trends in vaccine technology. This chapter does not include vaccines for parasitic disease [see Barbet, 1989], such as malaria, since such vaccines are not yet commercially available.

12.1 Antigen Cultivation

Microbial Cultivation

Bacterial Growth

As far as cultivation is concerned, the fundamental principles of cell growth are identical for the various types of bacterial vaccines. Detailed aspects of bacterial growth and metabolism can be found in Ingraham et al. [1983]. In the simplest whole-cell vaccines, obtaining cell mass is the objective of cultivation. Growth is an autocatalytic process where the cell duplicates by fission; growth is described by the differential equation,

$$\frac{dX}{dt} = \mu X_v, \tag{12.1}$$

where X is the total mass concentration of cells, X_v is the mass concentration of living, or viable cells, and μ is the specific growth rate in units of cells per cell time. In this equation, the growth rate of the cell is a function of the cells' physical and chemical environment:

$$\mu = f\left(\text{temperature, pH, dissolved oxygen, } C_1, C_2, C_3, \ldots, C_n\right). \tag{12.2}$$

All bacteria have a narrow range of temperatures permissive to growth; for human pathogens, the optimal temperature is usually 37°C; however, for attenuated strains, the optimal temperature may be purposefully lowered by mutation. At the end of culture, heat in excess of 50°C is sometimes used to kill pathogenic bacteria. Dissolved oxygen concentration is usually critical, since the organism will either require it or be inhibited by it. Cultivation of *Clostridium tetani*, for example, is conducted completely anaerobically.

In Eq. (12.2), C_i is the concentration of nutrient i presented to the cells. Nutrient conditions profoundly affect cell growth rate and the ultimate cell mass concentration achievable. At a minimum, the cells must be presented with a carbon source, a nitrogen source, and various elements in trace quantities. The carbon source, usually a carbohydrate such as glucose, is used for energy metabolism and as a precursor for various anabolites, small chemicals used to build the cell. The nitrogen source, which can be ammonia, or a complex amino acid mix such as a protein hydrolysate, also can be used to produce energy but is mainly present as a precursor for amino acid anabolites for protein production. The elements K, Mg, Ca, Fe, Mn, Mo, Co, Cu, and Zn are cofactors required in trace quantities for enzymatic reactions in the cell. Inorganic phosphorus and sulfur are incorporated into proteins, polysaccharides, and polynucleotides.

Historically, bacterial vaccines have been grown in a batchwise fashion on complex and ill-defined nutrient mixtures, which usually include animal and/or vegetable protein digests. An example is *Bordetella pertussis* cultivation on Wheeler and Cohen medium [World Health Organization, 1977c], which contains corn starch, an acid digest of bovine milk, casein, and a dialyzed lysate of brewer's yeast cells. The starch serves as the carbon source. The casein digest provides amino acids as a nitrogen source. The yeast extract provides amino acids and vitamins, as well as some trace elements, nucleotides, and carbohydrates. Such complex media components make it difficult to predict the fermentation performance, since their atomic and molecular compositions are not easily analyzed, and since they may contain hidden inhibitors of cell growth or antigen production. More recently, defined media have been used to better reproduce cell growth and antigen yield. In a defined medium, pure sugars, amino acids, vitamins, and other antibodies are used. When a single nutrient concentration limits cell growth, the growth rate can be described by Monod kinetics:

$$\mu = \frac{\mu_{\max} C_i}{C_i + K}. \tag{12.3}$$

Here $\mu_{\max}$ is the maximum specific rate, and K is the Monod constant for growth on the nutrient, an empirically determined number. It is seen from this equation that at high nutrient concentrations, the cell grows at maximum rate, as if the substrate were not limiting. At low concentration as the nutrient becomes limiting, the growth rate drops to zero, and the culture ends.

Antigen Production

Cultivation of microbes for subunit vaccines is similar to that of whole-cell vaccines, but here one is concerned with the production of the antigenic protein, polysaccharide, or antigen-encoding polynucleotide as opposed to the whole cell; thus cell growth is not necessarily the main objective. The goal is to maintain the proper environmental and nutritional factors for production of the desired product. Similar to Eqs. (12.1) and (12.2) above, one can describe antigen production by

$$\frac{dP}{dt} = q_p X_v - k_d P, \text{ where } q_p \text{ and}$$

$$k_d = f\left(\text{temperature pH, dissolved O}_2, C_1, \ldots, C_n\right), \tag{12.4}$$

where P is the product antigen concentration, q_p is the cell-specific productivity, and k_d is a degradation rate constant. For example, the production of pertussis toxin occurs best at slightly alkaline pH, about 8.0 [Jagicza et al., 1986]. *Corynebacterium diphtheriae* toxin production is affected by the concentration of iron [Reylveld, 1980]. Nutrient effects on degradation can be found for hepatitis B virus vaccine, whose production is actually a microbial cultivation. Here, recombinant DNA inserted into *Saccharomyces cerevisiae* yeast cells produces the hepatitis B surface antigen protein (HBsAg), which spontaneously assembles into virus-like particles (about 100 protein monomers per particle) within the cell. Prolonged starvation of the yeast cells can cause production of cellular proteases that degrade the antigen protein to provide nutrition. Consequently, maximum production of antigen is accomplished by not allowing the yeast culture to attain maximum cell mass but by harvesting the culture prior to nutrient depletion.

Cultivation Technology

Cultivation vessels for bacterial vaccines were initially bottles containing stagnant liquid or agar-gelled medium, where the bacteria grew at the liquid or agar surface. These typically resulted in low concentrations of the bacteria due to the limited penetration depth of diffusion-supplied oxygen and/or due to the limited diffusibility of nutrients through the stagnant liquid or agar. Despite the low mass, growth of the bacteria at the gas-liquid interface can result in cell differentiation (pellicle formation) and improved production of antigen (increased q_p). For the past several decades, however, glass and stainless steel fermentors have been used to increase production scale and productivity. Fermentors mix a liquid culture medium via an impeller, thus achieving quicker growth and higher cell mass concentrations than would be achievable by diffusive supply of nutrients. For aerobic microbe cultivations, fermenters oxygenate the culture by bubbling air directly into the medium. This increases the oxygen supply rate dramatically over diffusion supply. Due to the low solubility of oxygen in water, oxygen must be continuously supplied to avoid limitation of this nutrient.

Virus Cultivation

Virus cultivation is more complex than bacterial cultivation because viruses are, by themselves, nonreplicating. Virus must be grown on a host cell substrate, which can be animal tissue, embryo, or *ex vivo* cells; the host substrate determines the cultivation technology. In the United States, only Japanese encephalitis virus vaccine is still produced from infected mature animals. Worldwide, many vaccines are produced in chicken embryos, an inexpensive substrate. The remainder of vaccines are produced from *ex vivo* cultivated animal cells. Some virus-like particle vaccines are made by recombinant DNA techniques in either microbial or animal cells, e.g., hepatitis B virus vaccine, which is made in yeast as mentioned above, or in Chinese hamster ovary cells. An interesting synopsis of the development of rabies vaccine technology from Pasteur's use of animal tissues to modern use of *ex vivo* cells can be found in Sureau [1987].

In Vivo Virus Cultivation

Virus cultivation *in vivo* is straightforward, since relatively little control can be exercised on the host tissue. Virus is simply inoculated into the organ, and after incubation, the infected organ is harvested. Influenza virus is the prototypical *in vivo* vaccine; the virus is inoculated into the allantoic sac of 9- to 11-day-old fertilized chicken eggs. The eggs are incubated at about 33°C for 2 to 3 days, candled for viability and lack of contamination from the inoculation, and then the allantoic fluid is harvested. The process of inoculating, incubating, candling, and harvesting hundreds of thousands of eggs can be highly automated [Metzgar and Newhart, 1977].

Ex Vivo Virus Cultivation

Use of *ex vivo* cell substrates is the most recent technique in vaccine cultivation. In the case of measles and mumps vaccines, the cell substrate is chicken embryo cells that have been generated by trypsin enzyme treatment that dissociates the embryonic cells. Similarly, some rabies vaccines use cells derived from fetal rhesus monkey kidneys. Since the 1960s, cell lines have been generated that can be characterized, banked, and cryopreserved. Cryopreserved cells have been adopted to ensure reproducibility and freedom from contaminating viruses and microorganisms and to bypass the ethical problem of extensive use of animal tissues. Examples of commonly used cell lines are WI-38 and MRC-5, both human embryonic lung cells, and Vero, an African green monkey kidney cell line.

Ex vivo cells must be cultivated in a bioreactor, in liquid nutrient medium, with aeration. The principles of cell growth are the same as for bacterial cells described above, with some important additions. First, *ex vivo* animal cells cannot synthesize the range of metabolites and hormones necessary for survival and growth and must be provided with these compounds. Second, virus production is usually fatal to the host cells. Cells also become more fragile during infection, a process that is to an extent decoupled from cell growth. Virus growth and degradation and cell death kinetics can influence the choice of process or bioreactor. Third, all *ex vivo* cells used for human vaccine manufacture require a surface to adhere to in order to grow and function properly. Finally, animal cells lack the cross-linked, rigid polysaccharide cell wall that gives physical protection to microorganisms. These last three factors combine to necessitate specialized bioreactors.

Cell Growth.   Supplying cells with nutrients and growth factors is accomplished by growing the cells in a complex yet defined medium, which is supplemented with 2% to 10% (v/v) animal blood serum. The complex medium will typically contain glucose and L-glutamine as energy sources, some or all of the 20 predominant L-amino acids, vitamins, nucleotides, salts, and trace elements. For polio virus, productivity is a function of energy source availability [Eagle and Habel, 1956] and may depend on other nutrients. For polio and other viruses, the presence of the divalent cations Ca^{2+} and Mg^{2+} promotes viral attachment and entry into cells and stabilizes the virus particles. Serum provides growth-promoting hormones, lipids and cholesterol, surface attachment-promoting proteins such as fibronectin, and a host of other functions. The serum used is unusually of fetal bovine origin, since this source is particularly rich in growth-promoting hormones and contains low levels of antibodies that could neutralize virus. After cell growth, and before infection, the medium is usually changed to serum-free or low-serum medium. This is done both to avoid virus neutralization and to reduce the bovine protein impurities in the harvested virus. These proteins are immunogenic and can be difficult to purify away, especially for live-virus vaccines.

Virus Production Kinetics.   An intriguing aspect of *ex vivo* virus cultivation is that each process depends on whether the virus remains cell-associated, is secreted, or lyses the host cells and whether the virus is stable in culture. With few exceptions, a cell produces a finite amount of virus before dying, as opposed to producing the virus persistently. This is because the host cell protein and DNA/RNA synthesis organelles are usually commandeered by virus synthesis, and the host cannot produce its own proteins, DNA, or RNA. The goal then becomes to maximize cell concentration, as outlined above for bacterial vaccines, while maintaining the cells in a competent state to produce virus. Since infection is transient and often rapid, cell-specific virus productivity is not constant, and productivity is usually correlated with the cell state at inoculation. Specific productivity can be a function of the cell growth rate, since this determines the available protein and nucleic acid synthetic capacity. Although virus production can be nutrient-limited, good nutrition is usually ensured by the medium exchange at inoculation mentioned above. For viruses that infect cells slowly, nutrition is supplied by exchanging the medium several times, batchwise or by continuous perfusion.

For many viruses, the degradation term in Eq. (12.4) can be significant. This can be due to inherent thermal instability of the virus, oxidation, or proteases released from lysed cells. For an unstable virus, obtaining a synchronized infection can be key to maximizing titers. Here, the multiplicity of infection (MOI), the number of virus inoculated per cell, is an important parameter. An MOI greater than unity results in most cells being infected at once, giving the maximum net virus production rate.

Cultivation Technology.   The fragility of animal cells during infection and the surface attachment requirement create a requirement for special reactors [Prokop and Rosenberg, 1989]. To an extent, reactor choice and productivity are determined by the reactor surface area. Small, simple, and uncontrolled vessels, such as a flat-sided T-flask or Roux bottle, made of polystyrene or glass, can be used for small-scale culture. With these flasks the medium is stagnant, so fragile infected cells are not exposed to fluid motion forces during culture. Like bacterial cultures, productivity can be limited due to the slow diffusion of nutrients and oxygen. To obtain larger surface areas, roller bottles or Cell Factories are used, but like egg embryo culture, robotic automation is required to substantially increase the scale of production. Even larger culture scales ($\geq$50 liters) are accommodated in glass or stainless steel bioreactors, which are actively supplied with oxygen and pH controlled. The growth surface is typically supplied as a fixed bed of plates, spheres, or fibers or, alternately, as a dispersion of about 200-μm spherical particles known as *microcarriers* [Reuveny, 1990]. Microcarrier bioreactors are used for polio and rabies production [Montagnon et al., 1984]. The stirred-tank microcarrier reactors are similar to bacterial fermentors but are operated at much lower stirring speeds and gas sparging rates so as to minimize damage to the fragile cells. Packed-bed reactors are used for hepatitis A virus vaccine production [Aboud et al., 1994]. These types of reactors can give superior performance for highly lytic viruses because they subject the cells to much lower fluid mechanical forces. Design considerations for animal cell reactors may be found in Aunins and Henzler [1993].

12.2 Downstream Processing

Following cultivation, the antigen is recovered, isolated in crude form, further purified, and/or inactivated to give the unformulated product; these steps are referred to collectively as the *downstream process*. The complexity of a downstream process varies greatly depending on whether the antigen is the whole organism, a semipurified subunit, or a highly purified subunit. Although the sequence and combination of steps are unique for each vaccine, there is a general method to purification. The first steps reduce the volume of working material and provide crude separation from contaminants. Later steps typically resolve molecules more powerfully but are limited to relatively clean feed streams. Some manufacturing steps are classic small laboratory techniques, because many vaccine antigens are quite potent, requiring only micrograms of material; historically, manufacturing scale-up has not been a critical issue.

Purification Principles

Recovery

Recovery steps achieve concentration and liberation of the antigen. The first recovery step consists of separating the cells and/or virus from the fermentation broth or culture medium. The objective is to capture and concentrate the cells for cell-associated antigens or to clarify the medium of particulates for extracellular antigens. In the first case, the particle separation simultaneously concentrates the antigen. The two methods used are centrifugation and filtration. Batch volume and feed solids content determine whether centrifugation or filtration is appropriate; guidelines for centrifugation are given by Datar and Rosen [1993] and Atkinson and Mavituna [1991]. Filtration is used increasingly as filter materials science improves to give filters that do not bind antigen. Filtration can either be dead-end or cross-flow. Dead-end filters are usually fibrous depth filters and are used when the particulate load is low and the antigen is extracellular. In cross-flow filtration, the particles are retained by a microporous ($\geq 0.1\ \mu m$) or ultrafiltration ($\leq 0.1\ \mu m$) membrane. The feed stream is circulated tangential to the membrane surface at high velocity to keep the cells and other particulates from forming a filter cake on the membrane [Hanisch, 1986].

If the desired antigen is subcellular and cell-associated, the next recovery step is likely to be cell lysis. This can be accomplished by high-pressure valve homogenization, bead mills, or chemical lysis with detergent or chaotropic agents, to name only a few techniques. The homogenate or cell lysate may subsequently be clarified, again using either centrifugation or membrane filtration.

Isolation

Isolation is conducted to achieve crude fractionation from contaminants that are unlike the antigen; precipitation and extraction are often used here. These techniques rely on large differences in charge or solubility between antigen and contaminants. Prior to isolation, the recovered process stream may be treated enzymatically with nucleases, proteases, or lipases to remove DNA/RNA, protein, and lipids, the major macromolecular contaminants. Nonionic detergents such as Tween and Triton can serve a similar function to separate components, provided they do not denature the antigen. Subsequent purification steps must be designed to remove the enzyme(s) or detergent, however.

Ammonium sulfate salt precipitation is a classic method used to concentrate and partially purify various proteins, e.g., diphtheria and tetanus toxins. Alcohol precipitation is effective for separating polysaccharides from proteins. Cohn cold alcohol precipitation is a classic technique used to fractionate blood serum for antibody isolation. Both techniques concentrate antigen for further treatment.

Liquid-liquid extraction, either aqueous-organic or aqueous-aqueous, is another isolation technique suitable for vaccine purification. In a two-phase aqueous polymer system, the separation is based on the selective partitioning of the product from an aqueous liquid phase containing, e.g., polyethylene glycol (PEG), into a second, immiscible aqueous liquid phase that contains another polymer, e.g., dextran, or containing a salt [see Kelley and Hatton, 1992].

Final Purification

Further purification of the vaccine product is to remove contaminants that have properties closely resembling the antigen. These sophisticated techniques resolve molecules with small differences in charge, hydrophobicity, density, or size.

Density-gradient centrifugation, although not readily scalable, is a popular technique for final purification of viruses [Polson, 1993]. Either rate-zonal centrifugation, where the separation is based on differences in sedimentation rate, or isopycnic equilibrium centrifugation, where the separation is based solely on density differences, is used. Further details on these techniques can be found in Dobrota and Hinton [1992].

Finally, different types of chromatography are employed to manufacture highly pure vaccines; principles can be found in Janson and Ryden [1989]. Ion-exchange chromatography (IEC) is based on differences in overall charge and distribution of charge on the components. Hydrophobic-interaction chromatography (HIC) exploits differences in hydrophobicity. Affinity chromatography is based on specific stereochemical interactions common only to the antigen and the ligand. Size-exclusion chromatography (SEC) separates on the basis of size and shape. Often, multiple chromatographic steps are used, since the separation mechanisms are complementary. IEC and HIC can sometimes be used early in the process to gain substantial purification; SEC is typically used as a final polishing step. This technique, along with ultrafiltration, may be used to exchange buffers for formulation at the end of purification.

Inactivation

For nonattenuated whole organisms or for toxin antigens, the preparation must be inactivated to eliminate pathogenicity. This is accomplished by heat pasteurization, by cross-linking using formaldehyde or glutaraldehyde, or by alkylating using agents such as β-propiolactone. The agent is chosen for effectiveness without destruction of antigenicity. For whole organisms, the inactivation abolishes infectivity. For antigens such as the diphtheria and tetanus toxins, formaldehyde treatment removes the toxicity of the antigen itself as well as killing the organism. These detoxified antigens are known as *toxoids*.

The placement of inactivation in the process depends largely on safety issues. For pathogen cultures, inactivation traditionally has been immediately after cultivation to eliminate danger to manufacturing personnel. For inactivation with cross-linking agents, however, the step may be placed later in the process in order to minimize interference by contaminants that either foil inactivation of the organism or cause carryover of antigen-contaminant cross-linked entities that could cause safety problems, i.e., side-reactions in patients.

Purification Examples

Examples of purification processes are presented below, illustrating how the individual techniques are combined to create purification processes.

Bacterial Vaccines

Salmonella typhi Ty21, a vaccine for typhoid fever, and BCG (bacille Calmette-Guérin, a strain of *Mycobacterium bovis*) vaccine against *Myobacterium tuberculosis*, are the only vaccines licensed for human use based on live, attenuated bacteria. Downstream processing consists of collecting the cells by continuous-flow centrifugation or using cross-flow membrane filtration. For *M. bovis* that are grown in a liquid submerged fermenter culture, Tween 20 is added to keep the cells from aggregating, and the cultures are collected as above. Tween 20, however, has been found to decrease virulence. In contrast, if the BCG is grown in stagnant bottles as a surface pellicle, the downstream process consists of collecting the pellicle sheet, which is a moist cake, and then homogenizing the cake using a ball mill. Milling time is critical, since prolonged milling kills the cells and too little milling leaves clumps of bacteria in suspension.

Most current whooping cough vaccines are inactivated whole *B. pertussis*. The cells are harvested by centrifugation and then resuspended in buffer, which is the supernate in some cases. This is done because some of the filamentous hemagglutinin (FHA) and pertussis toxin (PT) antigens are released into the supernate. The cell concentrate is inactivated by mild heat and stored with thimerosal and/or formaldehyde. The inactivation process serves the dual purpose of killing the cells and inactivating the toxins.

C. diphtheria vaccine is typical of a crude protein toxoid vaccine. Here the 58 kDa toxin is the antigen, and it is converted to a toxoid with formaldehyde and crudely purified. The cells are first separated from the toxin by centrifugation. Sometimes the pathogen culture is inactivated with formaldehyde before centrifugation. The supernate is treated with formaldehyde to 0.75%, and it is stored for 4 to 6 weeks at 37°C to allow complete detoxification [Pappenheimer, 1984]. The toxoid is then concentrated by ultra-filtration and fractionated from contaminants by ammonium sulfate precipitation. During detoxification of crude material, reactions with formaldehyde lead to a variety of products. The toxin is internally cross-linked and also cross-linked to other toxins, beef peptones from the medium, and other medium proteins. Because detoxification creates a population of molecules containing antigen, the purity of this product is only about 60% to 70%.

Due to the cross-linking of impurities, improved processes have been developed to purify toxins before formaldehyde treatment. Purification by ammonium sulfate fractionation, followed by ion-exchange and/or size-exclusion chromatography, is capable of yielding diphtheria or tetanus toxins with purities ranging from 85% to 95%. The purified toxin is then treated with formaldehyde or glutaraldehyde [Relyveld and Ben-Efraim, 1983] to form the toxoid. Likewise, whole-cell pertussis vaccine is being replaced by subunit vaccines. Here, the pertussis toxin (PT) and the filamentous hemagglutinin (FHA) are purified from the supernate. These two antigens are isolated by ammonium sulfate precipitation, followed by sucrose density-gradient centrifugation to remove impurities such as endotoxin. The FHA and PT are then detoxified with formaldehyde [Sato et al., 1984].

Bacterial components other than proteins also have been developed for use as subunit vaccines. One class of bacterial vaccines is based on capsular polysaccharides. Polysaccharides from *Meningococcus*, *Pneumococcus*, and *Haemophilus influenzae* type b are used for vaccines against these organisms. After separating the cells, the polysaccharides are typically purified using a series of alcohol precipitations. As described below, these polysaccharides are not antigenic in infants. As a consequence, the polysaccharides are chemically cross-linked, or conjugated, to a highly purified antigenic protein carrier. After conjugation, there are purification steps to remove unreacted polysaccharide protein, and small-molecular-weight cross-linking reagents. Several manufacturers have introduced pediatric conjugate vaccines against *H. influenzae* type b (Hib-conjugate).

Viral Vaccines

Live viral vaccines have limited downstream processes. For secreted viruses such as measles, mumps, and rubella, cell debris is simply removed by filtration and the supernate frozen. In many cases it is necessary to process quickly and at refrigerated temperatures, since live virus can be unstable [Cryz and Gluck, 1990].

For cell-associated virus such as herpesviruses, e.g., varicella zoster (chickenpox virus), the cells are washed with a physiologic saline buffer to remove medium contaminants and are harvested. This can be done by placing them into a stabilizer formulation (see below) and then mechanically scraping the cells from the growth surface. Alternately, the cells can be harvested from the growth surface chemically or enzymatically. For the latter, the cells are centrifuged and resuspended in stabilizer medium. The virus is then liberated by disrupting the cells, usually by sonication, and the virus-containing supernate is clarified by dead-end filtration and frozen.

Early inactivated influenza virus vaccines contained relatively crude virus. The allantoic fluid was harvested, followed by formaldehyde inactivation of the whole virus, and adsorption to aluminum phosphate (see below), which may have provided some purification as well. The early vaccines, however, were associated with reactogenicity. For some current processes, the virus is purified by rate-zonal centrifugation, which effectively eliminates the contaminants from the allantoic fluid. The virus is then inactivated with formaldehyde or β-propiolactone, which preserves the antigenicity of both the hemagglutinin (HA) and neuraminidase (NA) antigens. Undesirable side reactions have been even further reduced with the introduction of *split vaccines*, where the virus particle is disrupted by an organic solvent or detergent and then inactivated. By further purifying the split vaccine using a method such as zonal centrifugation to separate the other virion components from the HA and NA antigens, an even more highly purified HA + NA vaccine is available. These antigens are considered to elicit protective antibodies against influenza [Tyrrell, 1976].

Recently, an extensive purification process was developed for hepatitis A vaccine, yielding a >90% pure product [Aboud et al., 1994]. The intracellular virus is released from the cells by Triton detergent lysis, followed by nuclease enzyme treatment for removal of RNA and DNA. The virus is concentrated and detergent removed by ion-exchange chromatography. PEG precipitation and chloroform solvent extraction purify away most of the cellular proteins, and final purification and polishing are achieved by ion-exchange and size-exclusion chromatography. The virus particle is then inactivated with formaldehyde. In this case, inactivation comes last for two reasons. First, the virus is attenuated, so there is no risk to process personnel. Second, placing the inactivation after the size-exclusion step ensures that there are no contaminants or virus aggregates that may cause incomplete inactivation.

The first hepatitis B virus vaccines were derived from human plasma [Hilleman, 1993]. The virus is a 22-nm-diameter particle, much larger than most biologic molecules. Isolation was achieved by ammonium sulfate or PEG precipitation, followed by rate zonal centrifugation and isopycnic banding to take advantage of the large particle size. The preparation was then treated with pepsin protease, urea, and formaldehyde or heat. The latter steps ensure inactivation of possible contaminant viruses from the blood serum. More recently, recombinant DNA–derived hepatitis B vaccines are expressed as an intracellular noninfectious particle in yeast and use a completely different purification process. Here, the emphasis is to remove the yeast host contaminants, particularly high levels of nucleic acids and polysaccharides. Details on the various manufacturing processes have been described by Sitrin et al. [1993].

Antibody Preparations

Antibody preparation starts from the plasma pool prepared by removing the cellular components of blood. Cold ethanol is added in increments to precipitate fractions of the blood proteins, and the precipitate containing IgG antibodies is collected. This is further redissolved and purified by ultrafiltration, which also exchanges the buffer to the stabilizer formulation. Sometimes ion-exchange chromatography is used for further purification. Although the plasma is screened for viral contamination prior to pooling, all three purification techniques remove some virus.

12.3 Formulation and Delivery

Successful vaccination requires both the development of a stable dosage form for in vitro storage and the proper delivery and presentation of the antigen to elicit a vigorous immune response *in vivo*. This is done by adjuvanting the vaccine and/or by formulating the adjuvanted antigen. An adjuvant is defined as an agent that enhances the immune response against an antigen. A formulation contains an antigen in a delivery vehicle designed to preserve the (adjuvenated) antigen and to deliver it to a specific target organ or over a desired time period. Despite adjuvanting and formulation efforts, most current vaccines require multiple doses to create immune memory.

Live Organisms

Live viruses and bacteria die relatively quickly in liquid solution (without an optimal environment) and are therefore usually stored in the frozen state. Preserving the infectivity of frozen live-organism vaccines is typically accomplished by lyophilization or freeze-drying. The freeze-drying process involves freezing the organism in the presence of stabilizers, followed by sublimation of both bulk water (primary drying) and more tightly bound water (secondary drying). The dehydration process reduces the conformational flexibility of the macromolecules, providing protection against thermal denaturation. Stabilizers also provide conformational stability and protect against other inactivating mechanisms such as amino acid deamidation, oxidation, and light-catalyzed reaction.

Final water content of the freeze-dried product is the most important parameter for the drying process. Although low water content enhances storage stability, overdrying inactivates biologic molecules, since removal of tightly bound water disrupts antigen conformation. Influenza virus suspensions have been shown to be more stable at 1.7% (w/w) water than either 0.4% to 1% or 2.1% to 3.2% [Greiff and Rightsel, 1968]. Other lyophilization parameters that must be optimized pertain to heat and mass transfer, including (1) the rate of freezing and sublimation, (2) vial location in the freeze-drier, and (3) the type of vial and stopper used to cap the vial. Rates of freezing and drying affect phase transitions and compositions, changing the viable organism yield on lyophilization and the degradation rate of the remaining viable organisms on storage.

Stabilizers are identified by trial-and-error screening and by examining the mechanisms of inactivation. They can be classified into four categories depending on their purpose: specific, nonspecific, competitive, and pharmaceutical. *Specific stabilizers* are ligands that naturally bind biologic macromolecules. For example, enzyme antigens are often stabilized by their natural substrates or closely related compounds. Antigen stabilizers for the liquid state also stabilize during freezing. *Nonspecific stabilizers* such as sugars, amino acids, and neutral salts stabilize proteins and virus structures via a variety of mechanisms. Sugars and polyols act as bound water substitutes, preserving conformational integrity without possessing the chemical reactivity of water. Buffer salts preserve optimal pH. *Competitive inhibitors* outcompete the organism or antigen for inactivating conditions, such as gas-liquid interfaces, oxygen, or trace-metal ions [Volkin and Klibanov, 1989]. Finally, *pharmaceutical stabilizers* may be added to preserve pharmaceutical elegance, i.e., to prevent collapse of the lyophilized powder during the drying cycle, which creates difficult redissolution. Large-molecular-weight polymers such as carbohydrates (dextrans or starch) or proteins such as albumin or gelatin are used for this purpose. For example, a buffered sorbitol-gelatin medium has been used successfully to preserve the infectivity of measles virus vaccine during lyophilized storage for several years at 2 to 8°C [Hilleman, 1989]. An example of live bacterium formulation to preserve activity on administration is typhoid fever vaccine, administered orally, *S. typhi* bacteria are lyophilized to a powder to preserve viability on the shelf, and the powder is encapsulated in gelatin to preserve bacterial viability when passing through the low-pH stomach. The gelatin capsule dissolves in the intestine to deliver the live bacteria.

Oral polio vaccine is an exception to the general rule of lyophilization, since polio virus is inherently quite stable relative to other viruses. It is formulated as a frozen liquid and can be used for a limited time after thawing [Melnick, 1984]. In the presence of specific stabilizers such as $MgCl_2$, extended 4°C stability can be obtained.

Subunit Antigens

Inactivated and/or purified viral and bacterial antigens inherently offer enhanced stability because whole-organism infectivity does not need to be preserved. However, these antigens are not as immunogenic as live organisms and thus are administered with an adjuvant. They are usually formulated as an aqueous liquid suspension or solution, although they can be lyophilized under the same principles as above. The major adjuvant recognized as safe for human use is alum. *Alum* is a general term referring to various hydrated aluminum salts; a discussion of the different alums can be found in Shirodkar et al. [1990]. Vaccines can be formulated with alum adjuvants by two distinct methods: adsorption to performed aluminum precipitates or precipitation of aluminum salts in the presence of the antigen, thus adsorbing and entrapping the antigen. Alum's adjuvant activity is classically believed to be a "depot" effect, slowly delivering antigen over time *in vivo*. In addition, alum particles are believed to be phagocytized by macrophages.

Alum properties vary depending on the salt used. Adjuvants labeled aluminum hydroxide are actually aluminum oxyhydroxide, AlO(OH). This material is crystalline, has a fibrous morphology, and has a positive surface charge at neutral pH. In contrast, aluminum phosphate adjuvants are networks of platelike particles of amorphous aluminum hydroxyphosphate and possess a negative surface charge at neutral pH. Finally, alum coprecipitate vaccines are prepared by mixing an acidic alum solution of $KAl(SO_4)_2 \cdot 12H_2O$ with an antigen solution buffered at neutral pH, sometimes actively pH-controlled with base. At neutral pH, the aluminum forms a precipitate, entrapping and adsorbing the antigen. The composition and physical properties of this alum vary with processing conditions and the buffer anions. In general, an amorphous aluminum hydroxy(buffer anion)sulfate material is formed.

Process parameters must be optimized for each antigen to ensure proper adsorption and storage stability. First, since antigen adsorption isotherms are a function of the antigen's isoelectric point and the type of alum used [Seeber et al., 1991], the proper alum and adsorption pH must be chosen. Second, the buffer ions in solution can affect the physical properties of alum over time, resulting in changes in solution pH and antigen adsorption. Finally, heat sterilization of alum solutions and precipitates prior to antigen adsorption can alter their properties. Alum is used to adjuvant virtually all the existing inactivated or formaldehyde-treated vaccines, as well as purified subunit vaccines such as HBsAg and Hib-conjugate vaccines. The exception is for some bacterial polysaccharide vaccines and for new vaccines under development (see below).

An interesting vaccine development challenge was encountered with Hib-conjugate pediatric vaccines, which consist of purified capsular polysaccharides. Although purified, unadjuvanted polysaccharide is used in adults, it is not sufficiently immunogenic in children under age 2, the population is greatest risk [Ellis, 1992; Howard, 1992]. Chemical conjugation, or cross-linking, of the PRP polysaccharide to an antigenic protein adjuvant elicits T-helper cell activation, resulting in higher antibody production. Variations in conjugation chemistry and protein carriers have been developed; example proteins are the diphtheria toxoid (CRM 197), tetanus toxoid, and the outer membrane protein complex of *N. meningitidis* [Ellis, 1992; Howard, 1992]. The conjugated polysaccharide is sometimes adsorbed to alum for further adjuvant action.

12.4 Future Trends

The reader will have noted that many production aspects for existing vaccines are archaic. This is so because most vaccines were developed before the biotechnology revolution, which is creating a generation of highly purified and better-characterized subunit vaccines. As such, for older vaccines "the process defines the product," and process improvements cannot readily be incorporated into these poorly characterized vaccines without extensive new clinical trials. With improved scientific capabilities, we can understand the effects of process changes on the physicochemical properties of new vaccines and on their behavior *in vivo*.

Vaccine Cultivation

Future cultivation methods will resemble existing methods of microbial and virus culture. Ill-defined medium components and cells will be replaced to enhance reproducibility in production. For bacterial and *ex vivo* cultivated virus, analytical advances will make monitoring the environment and nutritional status of the culture more ubiquitous. However, the major changes will be in novel product types — single-molecule subunit antigens, virus-like particles, monoclonal antibodies, and gene-therapy vaccines, each of which will incorporate novel processes.

Newer subunit vaccine antigens will be cultivated via recombinant DNA in microbial or animal cells. Several virus-like particle vaccines are under development using recombinant baculovirus (nuclear polyhedrosis virus) to infect insect cells (*spodoptera frugipeeda* or *trichoplusia ni*). Like the hepatitis B vaccine, the viral antigens spontaneously assemble into a noninfectious capsid within the cell. Although the metabolic pathways of insect cells differ from vertebrates, cultivation principles are similar. Insect cells do not require surface attachment and are grown much like bacteria. However, they also lack a cell wall and are larger and hence more fragile than vertebrate cells.

Passive antibody vaccines have been prepared up to now from human blood serum. Consequently, there has been no need for cultivation methods beyond vaccination and conventional harvest of antibody-containing blood from donors. Due to safety concerns over using human blood, passive vaccines will likely be monoclonal antibodies or cocktails thereof prepared in vitro by the cultivation of hybridoma or myeloma cell lines. This approach is under investigation for anti-HIV-1 antibodies [Emini et al., 1992]. Cultivation of these cell lines involves the same principles of animal cell cultivation as described above, with the exception that hybridomas can be less fastidious in nutritional requirements, and they do not require surface attachment for growth. These features will allow for defined serum-free media and simpler cultivation vessels and procedures.

For the gene-therapy approach, the patient actually produces the antigen. A DNA polynucleotide encoding protein antigen(s) is injected intramuscularly into the patient. The muscle absorbs the DNA and produces the antigen, thereby eliciting an immune response [Ulmer et al., 1993]. For cultivation, production of the DNA plasmid is the objective, which can be done efficiently by bacteria such as *Escherichia coli*. Such vaccines are not sufficiently far along in development to generalize the factors that influence their production; however, it is expected that producer cells and process conditions that favor high cell mass, DNA replication, and DNA stability will be important. A potential beauty of this vacci-nation approach is that for cultivation, purification, and formulation, many vaccines can conceivably be made by identical processes, since the plasmids are inactive within the bacterium and possess roughly the same nucleotide composition.

Downstream Processing

Future vaccines will be more highly purified in order to minimize side effects, and future improvements will be to assist this goal. The use of chemically defined culture media will impact favorably on down-stream processing by providing a cleaner feedstock. Advances in filtration membranes and in chromato-graphic support binding capacity and throughput will improve ease of purification. Affinity purification methods that rely on specific "lock-and-key" interactions between a chromatographic support and the antigen will see greater use as well. Techniques amenable to larger scales will be more important to meet increased market demands and to reduce manufacturing costs. HPLC and other analytical techniques will provide greater process monitoring and control throughout purification.

As seen during the evolution of diphtheria and tetanus toxoid vaccines, the trend will be to purify toxins prior to inactivation to reduce their cross-linking with other impurities. New inactivating agents such as hydrogen peroxide and ethyl dimethylaminopropyl carbodiimide have been investigated for pertussis toxin, which do not have problems of cross-linking or reversion of the toxoid to toxin status.

Molecular biology is likely to have an even greater impact on purification. Molecular cloning of proteins allows the addition of amino acid sequences that can facilitate purification, e.g., polyhistidine or poly-alanine tails for metal ion, or ion-exchange chromatography. Recent efforts also have employed genetic manipulation to inactivate toxins, eliminating the need for the chemical treatment step.

Vaccine Adjuvants and Formulation

Many new subunit antigens lack the inherent immunogenicity found in the natural organism, thereby creating the need for better adjuvants. Concomitantly, the practical problem of enhancing worldwide immunization coverage has stimulated development of single-shot vaccine formulations in which booster doses are unnecessary. Thus future vaccine delivery systems will aim at reducing the number of doses via controlled antigen release and will increase vaccine efficacy by improving the mechanism of antigen presentation (i.e., controlled release of antigen over time or directing of antigen to specific antigen-presenting cells). Major efforts are also being made to combine antigens into single-shot vaccines to improve immunization rates for infants, who currently receive up to 15 injections during the first 2 years of life. Coadministration of antigens presents unique challenges to formulation as well.

Recent advances in the understanding of *in vivo* antigen presentation to the immune system has generated considerable interest in developing novel vaccine adjuvants. The efficacy of an adjuvant is judged by its ability to stimulate specific antibody production and killer cell proliferation. Developments in biology now allow analysis of activity by the particular immune cells that are responsible for these processes. Examples of adjuvants currently under development include saponin detergents, muramyl dipeptides, and lipopolysaccharides (endotoxin), including lipid A derivatives. As well, cytokine growth factors that stimulate immune cells directly are under investigation.

Emulsion and liposome delivery vehicles are also being examined to enhance the presentation of antigen and adjuvant to the immune system [Edelman, 1992; Allison and Byars, 1992]. Controlled-release delivery systems are also being developed that encapsulate antigen inside a polymer-based solid microsphere. The size of the particles typically varies between 1 and 300 μm depending on the manufacturing process. Microspheres are prepared by first dissolving the biodegradable polymer in an organic solvent. The adjuvanted antigen, in aqueous solution or lyophilized powder form, is then emulsified into the solvent-polymer continuous phase. Microspheres are then formed by either solvent evaporation, phase-separation, or spray-drying, resulting in entrapment of antigen [Morris et al., 1994]. The most frequently employed biodegradable controlled-released delivery systems use FDA-approved poly(lactide-co-glycolide) copolymers (PLGA), which hydrolyze *in vivo* to nontoxic lactic and glycolic acid monomers. Degradation rate can be optimized by varying the microsphere size and the monomer ratio. Antigen stability during encapsulation and during *in vivo* release from the microspheres remains a challenge. Other challenges to manufacturing include encapsulation process reproducibility, minimizing antigen exposure to denaturing organic solvents, and ensuring sterility. Methods are being developed to address these issues, including the addition of stabilizers for processing purposes only. It should be noted that microsphere technology may permit vaccines to be targeted to specific cells; they can potentially be delivered orally or nasally to produce a mucosal immune response.

Other potential delivery technologies include liposomes and alginate polysaccharide and poly(dicarboxylatophenoxy)phosphazene polymers. The latter two form aqueous hydrogels in the presence of divalent cations [Khan et al., 1994]. Antigens can thus be entrapped under aqueous conditions with minimal processing by simply mixing antigen and soluble aqueous polymer and dripping the mixture into a solution of $CaCl_2$. The particles erode by Ca^{2+} loss, mechanical and chemical degradation, and macrophage attack. For alginate polymers, monomer composition also determines the polymer's immunogenicity, and thus the material can serve as both adjuvant and release vehicle.

For combination vaccines, storage and administration compatibility of the different antigens must be demonstrated. Live-organism vaccines are probably not compatible with purified antigens, since the former usually require lyophilization and the latter are liquid formulas. Within a class of vaccines, formulation is challenging. Whereas it is relatively straightforward to adjuvant and formulate a single antigen, combining antigens is more difficult because each has its own unique alum species, pH, buffer ion, and preservative optimum. Nevertheless, several combination vaccines have reached the market, and others are undergoing clinical trials.

12.5 Conclusions

Although vaccinology and manufacturing methods have come a considerable distance over the past 40 years, much more development will occur. There will be challenges for biotechnologists to arrive at safer, more effective vaccines for an ever-increasing number of antigen targets. If government interference and legal liability questions do not hamper innovation, vaccines will remain one of the most cost-effective and logical biomedical technologies of the next century, as disease is prevented rather than treated.

Challenges are also posed in bringing existing vaccines to technologically undeveloped nations, where they are needed most. This problem is almost exclusively dominated by the cost of vaccine manufacture and the reliability of distribution. Hence it is fertile ground for engineering improvements in vaccine production.

Defining Terms

Adjuvant: A chemical or biologic substance that enhances immune response against an antigen. Used here as a verb, the action of combining an antigen and an adjuvant.

Antigen: A macromolecule or assembly of macromolecules from a pathogenic organism that the immune system recognizes as foreign.

Attenuation: The process of mutating an organism so that it no longer causes disease.

Immunogen: A molecule or assembly of molecules with the ability to invoke an immune system response.

Pathogen: A disease-causing organism, either a virus, mycobacterium, or bacterium.

References

Aboud RA, Aunins JG, Buckland BC, et al. 1994. Hepatitis A Virus Vaccine. International patent application, publication number WO 94/03589, Feb. 17, 1994.

Allison AC, Byars NE. 1992. Immunologic adjuvants and their mode of action. In RW Ellis (ed), *Vaccines: New Approaches to Immunological Problems*, p 431. Reading, Mass., Butterworth-Heinemann.

Atkinson B, Mavituna F. 1991. *Biochemical Engineering and Biotechnology Handbook*, 2d ed. London, Macmillan.

Aunins JG, Henzler H-J. 1993. Aeration in cell culture bioreactors. In H-J Rehm et al. (eds), *Biotechnology*, 2d ed., vol 3, p 219. Weinheim, Germany, VCH Verlag.

Bachmayer H. 1976. Split and subunit vaccines. In P. Selby (ed), Influenza Virus, Vaccines, and Strategy, p 149. New York, Academic Press.

Barbet AF. 1989. Vaccines for parasitic infections. *Adv Vet Sci Comp Med* 33:345.

Cryz SJ, Reinhard G. 1990. Large-scale production of attenuated bacterial and viral vaccines. In GC Woodrow, MM Levine (eds), *New Generation Vaccines*, p 921. New York, Marcel Dekker.

Datar RV, Rosen C-G. 1993. Cell and cell debris removal: Centrifugation and crossflow filtration. In H-J Rehm et al. (eds), *Biotechnology*, 2d ed., vol 3, p 469. Weinheim, Germany, VCH Verlag.

Dobrota M, Hinton R. 1992. Conditions for density gradient separations. In D Rickwood (ed), *Preparative Centrifugation: A Practical Approach*, p 77. New York, Oxford U Press.

Eagle H, Habel K. 1956. The nutritional requirements for the propagation of poliomyelitis virus by the HeLa cell. *J Exp Med* 104:271.

Edelman R. 1992. An update on vaccine adjuvants in clinical trial. *AIDS Res Hum Retrovir* 8(8):1409.

Ellis RW. 1992. Vaccine development: Progression from target antigen to product. In JE Ciardi et al. (eds), *Genetically Engineered Vaccines*, p 263. New York, Plenum Press.

Emini EA, Schleif WA, Nunberg JH, et al. 1992. Prevention of HIV-1 infection in chimpanzees by gp120 V3 domain-specific monoclonal antibodies. *Nature* 355:728.

Greiff D, Rightsel WA. 1968. Stability of suspensions of influenza virus dried to different contents of residual moisture by sublimation in vacuo. *Appl Microbiol* 16(6):835.

Hanisch W. 1986. Cell harvesting. In WC McGregor (ed), *Membrane Separations in Biotechnology*, p 66. New York, Marcel Dekker.

Hewlett EL, Cherry JD. 1990. New and improved vaccines against pertussis. In GC Woodrow, MM Levine (eds), *New Generation Vaccines*, p 231. New York, Marcel Dekker.

Hilleman MR. 1989. Improving the heat stability of vaccines: Problems, needs and approaches. *Rev Infect Dis* 11(suppl 3):S613.

Hilleman MR. 1993. Plasma-derived hepatitis B vaccine: A breakthrough in preventive medicine. In R Ellis (ed), *Hepatitus B Vaccines in Clinical Practice*, p 17. New York, Marcel Dekker.

Howard AJ. 1992. Haemophilus influenzae type-b vaccines. *Br J Hosp Med* 48(1):44.

Ingraham JL, Maaløe O, Neidhardt FC. 1983. *Growth of the Bacterial Cell*. Sunderland, Mass, Sinauer.

Jagicza A, Balla P, Lendvai N, et al. 1986. Additional information for the continuous cultivation of *Bordetella* pertussis for the vaccine production in bioreactor. *Ann Immunol Hung* 26:89.

Janson J-C, Ryden L (eds). 1989. *Protein Purification Principles, High Resolution Methods, and Applications.* Weinheim, Germany, VCH Verlag.

Kelley BD, Hatton TA. 1993. Protein purification by liquid-liquid extraction. In H-J Rehm et al. (eds), *Biotechnology*, 2d ed, vol 3, p 594. Weinheim, Germany, VCH Verlag.

Khan MZI, Opdebeeck JP, Tucker IG. 1994. Immunopotentiation and delivery systems for antigens for single-step immunization: Recent trends and progress. *Pharmacol Res* 11(1):2.

Melnick JL. 1984. Live attenuated oral poliovirus vaccine. *Rev Infect Dis* 6(suppl 2):S323.

Metzgar DP, Newhart RH. 1977. U.S. patent no. 4,057,626, Nov. 78, 1977.

Montagnon B, Vincent-Falquet JC, Fanget B. 1984. Thousand litre scale microcarrier culture of vero cells for killed polio virus vaccine: Promising results. *Dev Biol Stand* 55:37.

Morris W, Steinhoff MC, Russell PK. 1994. Potential of polymer microencapsulation technology for vaccine innovation. *Vaccine* 12(1):5.

Pappenheimer AM. 1984. Diphtheria. In R Germanier (ed), *Bacterial Vaccines*, p 1. New York, Academic Press.

Polson A. 1993. *Virus Separation and Preparation.* New York, Marcel Dekker.

Prokop A, Rosenberg MZ. 1989. Bioreactor for mammalian cell culture. In A Fiechter (ed), *Advances in Biochemical Engineering, vol 39: Vertebrate Cell Culture II and Enzyme Technology*, p 29. Berlin, Springer-Verlag.

Rappuoli R. 1990. New and improved vaccines against diphtheria and tetanus. In GC Woodrow, MM Levine (eds), *New Generation Vaccines*, p 251. New York, Marcel Dekker.

Relyveld EH. 1980. Current developments in production and testing of tetanus and diphtheria vaccines. In A Mizrahi et al. (eds), *Progress in Clinical and Biological Research, vol 47: New Developments with Human and Veterinary Vaccines*, p 51. New York, Alan R Liss.

Relyveld EH, Ben-Efraim S. 1983. Preparation of vaccines by the action of glutaraldehyde on toxins, bacteria, viruses, allergens and cells. In SP Colowic, NO Kaplan (eds), *Methods in Enzymology*, vol 93, p 24. New York, Academic Press.

Reuveny S. 1990. Microcarrier culture systems. In AS Lubiniecki (ed), *In Large-Scale Mammalian Cell Culture Technology*, p 271. New York, Marcel Dekker.

Sato Y, Kimura M, Fukumi H. 1984. Development of a pertussis component vaccine in Japan. *Lancet* 1(8369):122.

Seeber SJ, White JL, Helm SL. 1991. Predicting the adsorption of proteins by aluminum-containing adjuvants. *Vaccine* 9:201.

Shirodkar S, Hutchinson RL, Perry DL, et al. 1990. Aluminum compounds used as adjuvants in vaccines. *Pharmacol Res* 7(12):1282.

Sitrin RD, Wampler DE, Ellis R. 1993. Survey of licensed hepatitis B vaccines and their product processes. In R Ellis (ed), *Hepatitus B Vaccines in Clinical Practice*, p 83. New York, Marcel Dekker.

Sureau P. 1987. Rabies vaccine production in animal cell cultures. In A Fiechter (ed), *Advances in Biochemical Engineering and Biotechnology*, vol 34, p 111. Berlin, Springer-Verlag.

Tyrrell DAJ. 1976. Inactivated whole virus vaccine. In P Selby (ed), *Influenza, Virus, Vaccines and Strategy*, p 137. New York, Academic Press.

Ulmer JB, Donnelly JJ, Parker SE, et al. 1993. Heterologous protection against influenza by injection of DNA encoding a viral protein. *Science* 259(5102):1745.

Volkin DB, Klibanov AM. 1989. Minimizing protein inactivation. In TE Creighton (ed), *Protein Function: A Practical Approach*, pp 1–12. Oxford, IRL Press.

Further Information

A detailed description of all the aspects of traditional bacterial vaccine manufacture may be found in the World Health Organization technical report series for the production of whole-cell pertussis, diphtheria, and tetanus toxoid vaccines:

World Health Organization. 1997a. BLG/UNDP/77.1 Rev. 1. Manual for the Production and Control of Vaccines: Diphtheria Toxoid.

World Health Organization. 1997b. BLG/UNDP/77.2 Rev. 1. Manual for the Production and Control of Vaccines: Tetanus Toxoid.

World Health Organization. 1997c. BLG/UNDP/77.3 Rev. 1. Manual for the Production and Control of Vaccines: Pertussis Vaccine.

A description of all the aspects of cell culture and viral vaccine manufacture may be found in Spier RE, Griffiths JB. 1985. *Animal Cell Biotechnology*, vols 1 to 3. London, Academic Press.

For a review of virology and virus characteristics, the reader is referred to Fields BN, Knipe DM (eds). 1990. *Virology*, 2d ed, vols 1 and 2. New York, Raven Press.

13

Gene Therapy

Joseph M. Le Doux
The Center for Engineering in Medicine, and Surgical Services, Massachusetts General Hospital, Harvard Medical School, and the Shriners Burns Hospital

Jeffrey R. Morgan
The Center for Engineering in Medicine, and Surgical Services, Massachusetts General Hospital, Harvard Medical School, and the Shriners Burns Hospital

Martin L. Yarmush
The Center for Engineering in Medicine, and Surgical Services, Massachusetts General Hospital, Harvard Medical School, and the Shriners Burns Hospital

Gene therapy, the transfer of genes into cells for a therapeutic effect, is a new approach to the treatment of disease. The first clinically applicable system for efficiently delivering genes into mammalian cells was developed in the early 1980s and was based on a genetically engineered **retrovirus**, which, as part of its lifecycle, stably integrates its genome into the target cell's chromosomal DNA. Using **recombinant** DNA technology perfected in the mid-1970s, investigators replaced the viral genes with therapeutic genes and the resulting recombinant retrovirus shuttled these genes into the target cells. The potential applications of gene therapy are far reaching (Table 13.1) since there are over 4000 known human genetic diseases (many of which have no viable treatment) and virtually every human disease is profoundly influenced by genetic factors [Anderson, 1992]. In one of the first clinical applications, gene therapy was used to treat patients with ADA (adenosine deaminase) deficiency, a genetic defect which causes severe combined immune deficiency (SCID) and death at an early age [Anderson, 1992]. The patient's lymphocytes were isolated and **transduced** (insertion of a foreign gene into the genome of a cell) by a recombinant retrovirus encoding a functional ADA gene. These transduced cells were expanded in culture then reinfused back to the patient. These protocols were conducted after an exhaustive peer review process that laid the groundwork for future gene therapy protocols. Successful treatment of other genetic diseases is likely to be achieved in the future [Levine and Friedmann, 1993; Roemer and Friedmann, 1992]. In addition to inherited diseases, other viable targets for gene therapy include more prevalent disorders that show a complex genetic dependence (i.e., cancer and heart disease) as well as infectious diseases (i.e., human immunodeficiency virus (HIV) and applications in tissue engineering [Anderson, 1992; Morgan and Yarmush, 1998].

13.1 Background

Gene therapy protocols conduct gene transfer in one of two settings; either *ex vivo* or *in vivo* [Mulligan, 1993]. For *ex vivo* gene therapy, target cells or tissue are removed from the patient, grown in culture, genetically modified and then reinfused or retransplanted into the patient [Ledley, 1993]. *Ex vivo* gene therapy is limited to those tissues which can be removed, cultured *in vitro* and returned to the patient and cannot be applied to many important target tissues and organs such as the lungs, brain and heart. For *in vivo* gene therapy, the gene transfer agent is delivered directly to the target tissue/organ, and gene

TABLE 13.1 Target Diseases for Gene Therapy

Target Disease	Target Tissues	Corrective Gene	Gene Delivery Systems Used
Inherited			
ADA deficiency	Hematopoietic cells	ADA	RV, RM
Alpha-1 antitrypsin deficiency	Fibroblasts	Alpha-1 antitrypsin	RV
	Hepatocytes		RV
	Lung epithelia cells		AV
	Peritoneal mesothelial cells		AV
Alzheimer's disease	Nervous system	Nerve growth factor	RV, AV, HSV
Cystic fibrosis	Lung epithelia cells	CFTR	AV,AAV, RM, L
Diabetes	Fibroblasts	Human insulin	RV
	Hepatocytes		L
Duchenne's muscular dystrophy	Muscle cells	Dystrophin	AV, DI
Familial hypercholesterolemia	Hepatocytes	LDL receptor	RV, AV, RM
Gaucher's disease	Hematopoietic cells	Glucocerebrosidase	RV
	Fibroblasts		RV
Growth hormone deficiency	Endothelial cells	Human growth hormone	RV
	Fibroblasts		TR
	Keratinocytes		RV
	Muscle cells		RV
Hemoglobinopathies	Hematopoietic cells	α or β-globin	RV, AAV
Hemophilia	Fibroblasts	Factor VIII, IX	RV, DI, L
	Keratinocytes		RV
	Hepatocytes		RV, AV, RM, L
	Muscle cells		RV
Leukocyte adhesion deficiency	Hematopoietic cells	CD-18	RV
Parkinson's disease	Nervous system	Tyrosine hydroxylase	RV, AV, HSV
Phenylketonuria	Hepatocytes	Phenylalanine hydroxylase	RV
Purine nucleoside phosphorylase deficiency	Fibroblasts	Purine nucleoside phosphorylase	RV
Urea cycle disorders	Hepatocytes	Ornithine transcarbamylase or arginosuccinate synthetase	AV
Acquired			
Cancer	Acute lymphoblastic leukemia	p53	RV
	Brain tumors	HSV thymidine kinase	RV
	Carcinoma	γ-interferon	TR
	Melanoma	Tumor necrosis factor	RV
	Retinoblastoma	Retinoblastoma gene	RV
Infectious diseases	HIV	Dominant negative Rev	RV
		TAR decoy	RV
		RRE decoy	TR
		Diptheria toxin A	RV
Cardiomyopathy	Muscle cells	(Used reporter gene)	DI
Emphysema	Lung epithelial cells	Alpha-1 antitrypsin	AV
Local thrombosis	Endothelial cells	Anti-clotting factors	RV
Vaccines	Muscle cells	Various	DI

*Note:*AAV = recombinant adeno-associated viruses
AV = recombinant adenoviruses
DI = direct injection
HSV = recombinant herpes simplex virus
L = lipofection
RM = receptor mediated
RV = recombinant retroviruses
TR = transfection

TABLE 13.2 Physical Characteristics of Recombinant Virions

Characteristic	Units	Recombinant Retroviruses	Recombinant Adenoviruses	Recombinant AAV
Genome type		ss RNA (2 per virion)	ds DNA	ss DNA
Genome size	Bases	8300	36000	4700
Genome MW	Daltons	3×10^6	$20\text{--}25 \times 10^6$	$1.2\text{--}1.8 \times 10^6$
Particle diameter	nm	90–147	65–80	20–24
Particle mass	Grams	3.6×10^{-16}	2.9×10^{-16}	1.0×10^{-17}
Composition				
DNA/RNA	%	2	13	26
Protein	%	62	87	74
Lipid	%	36	0	0
Density	g/cm^3 CsCl	1.15–1.16	1.33–1.35	1.39–1.42
Enveloped?	Yes/no	Yes	No	No
Shape		Spherical	Icosahedral	Icosahedral
Surface projections?	Yes/no	Yes	Yes	No
Number		~60–200	12	
Length	nm	5	25–30	
Max diameter	nm	8	4	
Virus titer	pfu/ml	$10^6\text{--}10^7$	$10^{10}\text{--}10^{12}$	$10^5\text{--}10^7$
Integration?		Yes, random	No, episomal	Yes, chromosome 19

transfer occurs in the patient rather than in the tissue culture dish [Mulligan, 1993]. Both strategies have inherent advantages and disadvantages and current research is evaluating which approach can best meet the needs of a particular disease.

Gene delivery systems can be classified as either viral or nonviral [Friedmann, 1989]. For viral gene transfer, one of several different types of viruses is engineered to deliver genes. Typically, viral genes are removed to prevent self-replication of the virus and to provide room for the insertion of one or more therapeutic genes that the recombinant virus will carry. To further ensure the safety of the recombinant viruses, specialized **packaging cell lines** have been developed to produce the recombinant viruses and minimize the production of infectious wild-type viruses. Some viruses are able to integrate the therapeutic genes into the target cell's nuclear DNA (retroviruses, adeno-associated viruses), whereas others are not (adenoviruses) (Table 13.2).

Non-viral gene transfer systems are based on a variety of technologies that employ physical/chemical means to deliver genes [Felgner and Rhodes, 1991]. These technologies include direct **plasmid** injection, bombardment with DNA coated microprojectiles, and DNA complexed with **liposomes**. Some nonviral transfection techniques are too inefficient (e.g., coprecipitation of DNA with calcium phosphate [Chen and Okayama, 1987], DNA complexed with DEAE-dextran [Pagano et al., 1967], electroporation [Neumann et al., 1982]) or laborious (e.g., microinjection of DNA [Capecchi, 1980]) for clinical use. Only those gene delivery systems (viral and nonviral) with potential for clinical application will be discussed in this chapter. The main features of these technologies (Table 13.3) will be described and specific examples of their applications highlighted.

13.2 Recombinant Retroviruses

Many of the approved clinical trials have utilized recombinant retroviruses for gene delivery. Retroviral particles contain two copies of identical RNA genomes that are wrapped in a protein coat and further encapsidated by a lipid bilayer membrane. The virus attaches to specific cell surface receptors via surface proteins that protrude from the viral membrane. The particle is then internalized and its genome is released into the cytoplasm, reverse transcribed from RNA to DNA, transported into the nucleus and then integrated into the cell's chromosomal DNA. The integrated viral genome has LTRs (long terminal repeats) at both ends, which encode the regulatory sequences that drive the expression of the viral genome [Weiss et al., 1982].

TABLE 13.3   Features of the Various Gene Transfer Systems

Features	Retrovirus	AAV	Adenovirus	Nonviral
Maximum transgene size	8 kb	4.7 kb	36 kb	$\geqslant$36 kb
Maximum concentration (vectors/ml)	~10^7	~10^{12}	~10^{12}	$\geqslant$36 kb
Transfers genes to quiescent cells	No/Yes*	Yes	Yes	Yes
Integrates transgene into target cell genome	Yes	Yes	No	No
Persistence of gene expression	wks–yrs	yrs	wks–mos	days–wks
Immunological problems	Few	None known	Extensive	None
Pre-existing host immunity	No	Yes	Yes	No
Stability	Poor	Good	Good	Good
Ease of large scale production	Difficult	Difficult	Easy	Easy
Safety concerns	Insertional mutagenesis	Inflammation toxicity	Inflammation toxicity	Toxicity

* Recombinant lentiviruses, such as human immunodeficiency virus, are capable of transducing quiescent cells.

Retroviruses used for gene transfer are derived from wild-type murine retroviruses. The recombinant viral particles are structurally identical to the wild type virus but carry a genetically engineered genome (retroviral vector), which encodes the therapeutic gene of interest. These recombinant viruses are incapable of self-replication but can infect and insert their genomes into a target cell's genome [Morgan et al., 1993].

Recombinant retroviruses, like all other recombinant viruses, are produced by a two-part system composed of a packaging cell line and a recombinant vector (Fig. 13.1) [Anderson, 1992; Levine and Friedmann, 1993]. The packaging cell line has been engineered to express all the structural viral genes (*gag, pol* and *env*) necessary for the formation of an infectious virus particle. *gag* encodes the capsid proteins and is necessary for encapsidation of the vector. *pol* encodes the enzymatic activities of the virus including reverse transcriptase and integrase. *env* encodes the surface proteins on the virus particle that are necessary for attachment to the target cell's receptors.

The retroviral vector is essentially the wild-type genome with all the viral genes removed. This vector encodes the transgene(s) and the regulatory sequences necessary for their expression as well as a special packaging sequence, (ψ), which is required for encapsidation of the genome into an infectious viral particle [Morgan et al., 1993]. To produce recombinant retrovirus particles, the retroviral vector is transfected into the packaging cell line. The structural proteins expressed by the packaging cell line recognize the packaging sequence on RNAs transcribed from the transfected vector and encapsidate them into an infectious virus particle that is subsequently exocytosed by the cell and released into the culture medium. This medium containing infectious recombinant viruses is harvested and used to transduce target cells.

Many different retroviral vector designs have been used (Fig. 13.2). A commonly used vector encodes two genes, one expressed from the LTR and the other from an internal promoter (Fig. 13.2c) [Miller, 1992]. Often, one gene expresses a therapeutic protein and the other a selectable marker that makes the modified cells resistant to selective media or a drug. This allows the investigator to establish a culture composed solely of transduced cells by growing them under the selective conditions. Several configurations are possible, but the optimum design is often dictated by the transgene(s) being expressed and the cell type to be transduced. Vector configuration is crucial for maximizing viral titer and transgene expression [Roemer and Friedmann, 1992].

As with all gene transfer technologies, there are advantages and disadvantages to the use of recombinant retroviruses. Retroviruses can only transduce dividing cells since integration requires passage of the target cells through mitosis [Roe et al., 1993]. This limits the use of recombinant retroviruses for *in vivo* gene therapy, since few normal cells are actively dividing. Recently, however, new retroviral vectors based on lentiviruses have been developed that are capable of transducing nondividing cells [Naldini et al., 1996]. These HIV based vectors were able to achieve stable gene transfer after injection into a nondividing tissue (brain) *in vivo*.

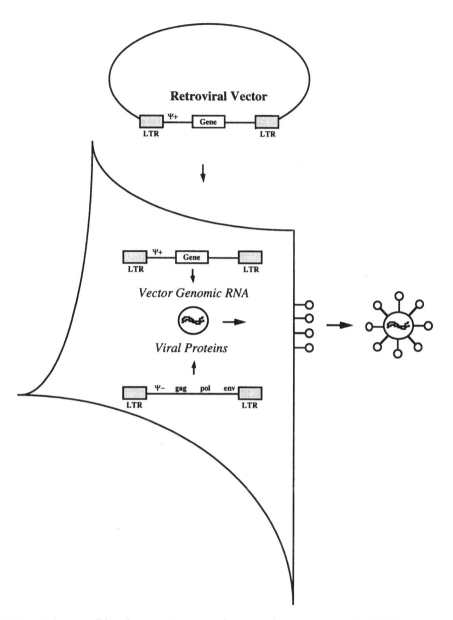

FIGURE 13.1   Packaging cell line for retrovirus. A simple retroviral vector composed of 2 LTR regions that flank sequences encoding the packaging sequence (Ψ) and a therapeutic gene. A packaging cell line is transfected with this vector. The packaging cell line expresses the three structural proteins necessary for formation of a virus particle (*gag, pol,* and *env*). These proteins recognize the packaging sequence on the vector and form an infectious virion around it. Infectious virions bud from the cell surface into the culture medium. The virus-laden culture medium is filtered to remove cell debris and then either immediately used to transduce target cells or the virions are purified and/or concentrated and frozen for later use.

Other drawbacks of retroviral vectors include: (a) a limitation to the size of inserted genes (<8 kilobases) [Roemer and Friedmann, 1992]; (b) the particles are unstable and lose activity at 37°C with a half-life of 5-7 hours [Kotani et al., 1994; Le Doux et al., 1999]; and (c) virus producer cell lines typically produce retrovirus in relatively low titers (10^5 to 10^7 infectious particles per ml) [Paul et al., 1993]. The viral titer is a function of several factors including the producer cell line, the type of transgene and the vector construction. Moreover, purification and concentration of retroviruses without loss of

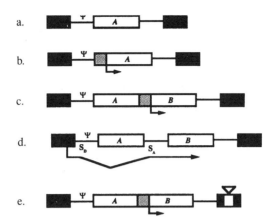

FIGURE 13.2 Standard retroviral vector designs. LTRs are shown as black boxes. S_D and S_A represent splice donor and acceptor sites, respectively. Ψ indicates the packaging signal sequence. Internal promoters are the stippled boxes and transcription start sites are indicated by arrows. (*a*) A single-gene vector. Transcription is driven from the LTR. (*b*) A single-gene vector expressed from an internal promoter. (*c*) A typical two-gene vector. One gene is the therapeutic gene, and the other is a selectable marker gene. Gene *A* is driven from the LTR, and gene *B* is expressed from the internal promoter. (*d*) A two-gene vector in which transcription is driven from the LTR. The efficiency of expression of gene *B* is a function of the efficiency of splicing. (*e*) A self-inactivating vector. A deletion in 3′ LTR is copied to the 5′ LTR during reverse transcription. This eliminates any transcription from the LTR. Internal promoters are still active. This vector reduces the chance of insertional activation of a proto-oncogene downstream from the 3′ LTR.

infectivity is difficult [Andreadis et al., 1999]. Standard techniques such as centrifugation, column filtration or ultrafiltration have failed [Le Doux et al., 1998; McGrath et al., 1978]. More recently, hollow fiber [Paul et al., 1993] and tangential flow filtration [Kotani et al., 1994] have been used with some success. A **chimeric retrovirus** in which the normal envelope proteins were replaced by VSV (Vesicular Stomatitis Virus) envelope proteins was readily concentrated by centrifugation [Burns et al., 1993].

Stability and level of expression has also been cited as a significant drawback for retrovirally transduced cells. Although long-term expression has been documented in some systems such as bone marrow, fibroblasts, hepatocytes and muscle cells, there have also been reports of transient expression in retrovirally transduced cells [Roemer and Friedmann, 1992]. This unstable expression has been attributed to several causes including methylation of the DNA and in some cases rearrangement or loss of the proviral sequences [Roemer and Friedmann, 1992].

The use of recombinant retroviruses has raised two major safety concerns [Roemer and Friedmann, 1992; Temin, 1990]. Older packaging cell lines occasionally produced replication competent virus by homologous recombination between the retroviral vector and the packaging cell line's retroviral sequences. New packaging cell lines have made the production of replication competent viruses essentially impossible [Danos and Mulligan, 1988]. The other safety concern was the possibility that the integration of the recombinant retrovirus would activate a proto-oncogene. The probability of this event is very low, and since several mutations are required for a cell to become cancerous, the risk of cellular transformation is extremely low and is typically outweighed by the potential therapeutic benefits [Temin, 1990].

13.3 Recombinant Adenoviruses

Recombinant adenoviruses have a number of properties that make them a useful alternative to recombinant retroviruses for human gene transfer. Recombinant adenoviruses are well-characterized, relatively easy to manipulate, can be grown and concentrated to very high titers (up to 10^{13} infectious particles/ml), are stable particles, and can transduce a wide variety of cell types [Crystal, 1995; Hitt et al., 1997]. Recombinant adenoviruses are also a relatively safe gene transfer system. Replication competent viruses, should they contaminate virus stocks or form by recombination with wild-type adenoviruses from latent infections, do not pose any significant risk since they elicit only a mild, self-limiting infection [Parks et al., 1996]. Furthermore, genes transferred by recombinant adenoviruses do not integrate, eliminating the risk of insertional mutagenesis of the chromosomal DNA of the target cell [Crystal, 1995]. Most important, recombinant adenoviruses efficiently transfer genes to nondividing, as well as dividing, cells that makes possible the *in vivo* transduction of tissues composed of fully differentiated or slowly dividing cells such as the liver and lung [Mulligan, 1993].

Adenoviruses consist of a large double stranded DNA genome (about 36 kilobase pairs long) packaged within a nonenveloped icosahedral capsid that is primarily composed of three virus-encoded proteins (hexon, penton base, and fiber proteins) [Horwitz, 1990]. The fiber proteins protrude from the surface of the virus and mediate its attachment to target cells via a high affinity interaction with cell surface receptors. The virus is then internalized into endosomal vesicles via specific interactions between the penton base proteins and α_v integrins [Wickham et al., 1993]. Adenoviruses escape these vesicles by an acid-induced endosomolytic activity and are transported to the nucleus, into which they enter via pores in the nuclear membrane [Greber et al., 1993].

The wild type adenovirus genome consists primarily of five early genes (E1-E5) each of which is expressed from their own promoters [Horwitz, 1990]. There are also five late genes (L1-L5), which are expressed from the major late promoter (MLP). The first generation of recombinant adenoviruses were based on a mutant adenovirus in which the E1 region (and in some cases the E3 region) was deleted. The E1 region is required for replication. Nevertheless, E1 minus mutants can be grown on specialized packaging cell lines (293 cells), which express the E1 gene and therefore provide the necessary functions for virus production [Levine and Friedmann, 1993].

To generate recombinant adenoviruses, a plasmid that contains the gene of interest, flanked by adenovirus sequences, is transfected into 293 cells in which adenoviruses that lack the E1 region are actively replicating (Fig. 13.3). The virus stock is screened for the rare recombinants in which the gene of interest has correctly recombined with the E1 minus mutant by homologous recombination. These recombinant virions are purified and grown to high titer on 293 cells [Morgan et al., 1994].

Recombinant adenoviruses have been successfully used for *in vivo* gene transfer in a number of animal models and in several clinical protocols, including those which delivered a functional copy of the cystic fibrosis transmembrane conductance regulator (CFTR) gene into the nasal epithelium and lungs of cystic fibrosis patients [Boucher and Knowles, 1993; Welsh, 1993; Wilmott and Whitsett, 1993]. Unfortunately, transgene expression was short-lived (5 to 10 days) [Verma and Somia, 1997]. Short-lived expression is a disadvantage for the treatment of genetic or chronic disorders because, in order to maintain the therapeutic effect of the transgene, recombinant adenoviruses would have to be administered to the patient repeatedly. Repeat administration often fails due to the presence of neutralizing antibodies that form in response to the first administration of the recombinant adenoviruses [Hitt et al., 1997]. Alternatively, short lived high expression from adenovirus vectors might be therapeutically useful for applications where transient expression is preferred. Such applications include the stimulation of angiogenesis by adenoviruses expressing vascular endothelial growth factor (VEGF) [Magovern et al., 1997; Morgan and Yarmush, 1998].

Short-lived gene expression is currently one of the most significant limitations of recombinant adenoviruses for use in human gene therapy. Although short-lived gene expression could be the result of vector cytotoxicity, promoter shutoff, or loss of the transgene DNA, most investigators believe it is primarily due to the elimination of transduced cells by transgene or adenovirus-specific cytotoxic T-lymphocytes [Dai et al., 1995; Engelhardt et al., 1994; Scaria et al., 1998; Zsengeller et al., 1995]. As a result, most efforts to increase the longevity of gene expression have focused on eliminating the immune rejection of cells transduced by recombinant adenoviruses. Two major strategies have been pursued. One strategy has been to construct recombinant adenoviruses that encode fewer immunogenic adenoviral proteins, such as by deleting portions of the E2 region or the E4 region. Such vectors have the added benefit of being able to accommodate the cloning of much larger inserts of therapeutic DNA (to as much as 36 kilobases) than was possible with the first-generation (E1 and E3 minus) recombinant adenoviruses which could accommodate only 7 kilobases of foreign DNA [Morsy et al., 1998; Parks et al., 1996].

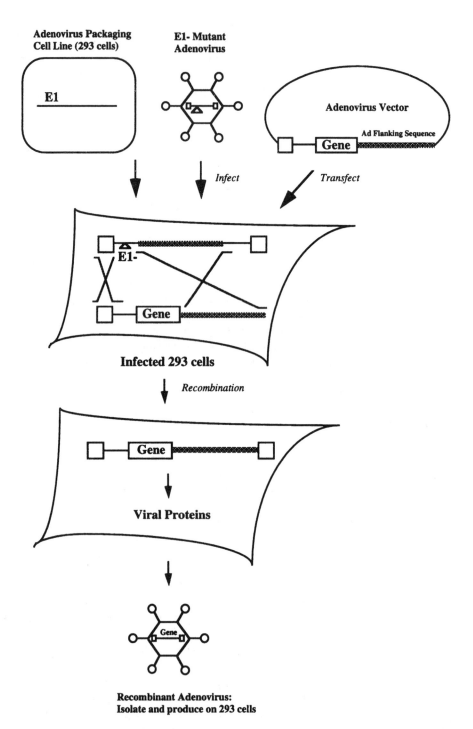

FIGURE 13.3 Isolation of recombinant adenovirus. A packaging cell line (293 cells), which expresses the E1 gene, is infected with an E1-minus mutant adenovirus. The adenovirus is derived from a plasmid encoding the wild-type adenovirus genome. The E1 region of the adenovirus genome is replaced with the therapeutic gene, and the resultant plasmid (the adenovirus vector) is transfected into the 293 cell line, which is infected by the mutant adenovirus. Since the therapeutic gene is flanked by adenoviral sequences, the mutant adenovirus genome and the adenovirus vector will occasionally undergo homologous recombination and form an infectious recombinant adenovirus whose genome encodes the therapeutic gene. These rare recombinants are isolated then grown on another 293 cell line.

The second strategy is to block the host immune response. For example, investigators have co-injected recombinant adenoviruses with immunomodulatory compounds (e.g., deoxyspergualin, IL-12, IFN-gamma) that transiently suppressed the immune system, eliminated the humoral response, and made possible repeat administration of the vector [Hitt et al., 1997]. Others have co-injected vectors that encode for immunosuppressive compounds that induce tolerance to recombinant adenoviruses (e.g., CTLA4Ig), or are developing recombinant adenoviruses whose backbones contain genes that encode for immuno-suppressive proteins (e.g., Ad5 E3 19-kDa protein or HSV ICP47) [Nakagawa et al., 1998; Wold and Gooding, 1989; York et al., 1994].

As a result of these and similar efforts, the persistance of gene expression *in vivo* has been extended to as long as 5 months [Kay et al., 1995; Scaria et al., 1998]. It remains to be determined, however, if these methods can eliminate the immune response against recombinant adenoviruses and their products, whether or not elimination of the immune response results in long-term gene expression, and if these methods will be effective in human gene therapy protocols.

13.4 Recombinant Adeno-Associated Viruses

Adeno-associated viruses (AAV) are another virus-based gene transfer system that has significant potential for use in human gene therapy. AAV are small, nonenveloped human parvoviruses that contain a single-stranded DNA genome (4.7 kilobases) and encode two genes required for replication, *rep* and *cap* [McCarty and Samulski, 1997]. AAV are stable, have a broad host range, can transduce dividing and nondividing cells, and do not cause any known human disease [Verma and Somia, 1997]. AAV require the presence of a helper virus (typically adenovirus) to replicate [Muzyczka, 1992]. When no helper virus is present, AAV do not replicate but instead tend to establish a latent infection in which they permanently integrate into the chromosomal DNA of the target cells [Inoue and Russell, 1998]. Wild-type AAV preferentially integrates into a specific site in chromosome 19 due to the interaction of the virus-encoded protein Rep with the host cell DNA [McCarty and Samulski, 1997]. Recombinant AAV, however, are Rep negative and as a result integrate randomly into the chromosomal DNA of the target cell.

To generate recombinant AAV, human 293 cells are transfected with the AAV vector, a plasmid that contains the therapeutic gene of interest flanked by the 145-bp AAV inverted terminal repeats (ITR) that are necessary for its encapsidation into a virus particle [Ferrari et al., 1997] (Fig. 13.4). The cells are also transfected with a helper plasmid, a plasmid that encodes for the virus proteins necessary for particle formation and replication (i.e., AAV *rep* and *cap* genes). The transfected cells are infected with wild-type adenovirus which supplies helper functions required for amplification of the AAV vector [Morgan and Yarmush, 1998]. Contaminating adenovirus is removed or inactivated by density gradient centrifugation and heat inactivation. Recombinant AAV generated by this and similar methods have been successfully used to achieve long-term expression of therapeutic proteins in a number of cell types and tissues, including in the lung, muscle, liver, central nervous system, retina, and cardiac myocytes [Bartlett et al., 1998; Flannery et al., 1997; Flotte et al., 1993; Maeda et al., 1998; Monahan et al., 1998; Xiao et al., 1998]. Therapeutic effects have been achieved in a number of model systems, including in a dog model of hemophilia and a mouse model for obesity and diabetes [Monahan et al., 1998; Murphy et al., 1997].

Despite these early successes, several technical issues must be addressed before recombinant AAV can be used for human gene therapy on a routine basis. For example, little is known about the conditions or factors which enhance recombinant AAV transduction efficiency *in vitro* or *in vivo* [McCarty and Samulski, 1997]. Nor is it well-understood what controls the efficiency with which recombinant AAV integrate into the chromosomal DNA of target cells or why the efficiency of integration is so low (<1%). Perhaps the most significant technical issue, however, is that the current methods for producing recombinant AAV (see above) are tedious, labor intensive, and not well-suited for producing clinical grade virus.

Current methods are not adequate for producing a clinical grade virus for several reasons. First, current methods do not produce enough virus particles to be useful for many gene therapies. For example, it has been estimated that 10^{14} recombinant viruses will be needed to achieve systemic production of therapeutic levels of proteins such as factor IX or erythropoietin [Inoue and Russell, 1998]. With current

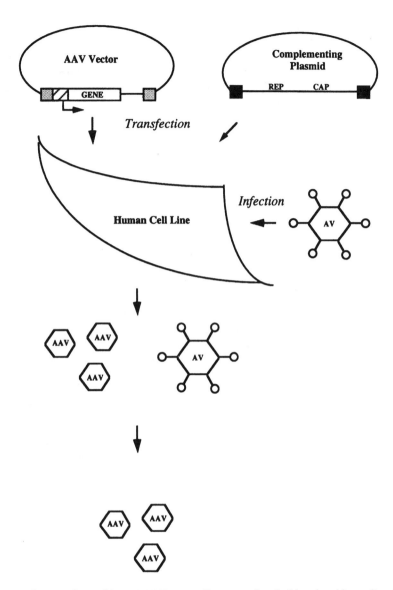

FIGURE 13.4 Production of recombinant AAV. Human cells are transfected with a plasmid encoding the therapeutic gene which is driven from a heterologous promoter (*cross-hatched box*) and flanked with AAV terminal repeats (*stippled boxes*). The cells are also transfected with a complementing plasmid encoding the AAV *rep* and *cap* genes, which cannot be packaged because they are not flanked by AAV terminal repeats. The *rep* and *cap* genes, whose products are required for particle formation, are flanked by adenovirus 5 terminal fragments (*black boxes*) that enhance their expression. The transfected cells are infected with wild-type adenovirus which supplies helper functions required for amplification of the AAV vector. The virus stock contains both AAV recombinant virus and adenovirus. The adenovirus is either separated by density gradient centrifugation or heat inactivated.

production methods, which yield 10^8 to 10^9 viruses per ml, about 100 to 1000 liters of virus would have to be produced for each treatment. Second, current methods require the use of helper adenovirus which is a significant disadvantage because their removal is tedious and labor intensive. In addition, there is the risk that residual contaminating adenovirus proteins will be present and stimulate the immune rejection of transduced cells [Xiao et al., 1998]. Finally, the use of transient transfection is not amenable to scale up and significantly increases the likelihood that replication competent AAV will be generated by recombinogenic events [Inoue and Russell, 1998].

To overcome these problems, investigators are working toward the development of stable packaging cell lines that produce recombinant AAV. Substantial increases in the yield of recombinant AAV have been achieved by increasing in the virus producer cell lines the number of copies of the AAV vector or of the helper plasmid. For example, one group developed an AAV packaging cell line that contained integrated helper and vector constructs that were linked to the simian virus 40 replication origin [Inoue and Russell, 1998]. These packaging cells could be induced to express SV40 T antigen, which then amplified the helper and vector constructs by 4 to 10 fold, resulting in 5- to 20-fold increases in recombinant AAV production. Progress has also been made toward the elimination of the need for helper adenovirus. For example, adenovirus miniplasmids have been constructed that provide all of the adenovirus helper functions but do not support the production of adenovirus proteins or viruses [Xiao et al., 1998]. Use of adenovirus miniplasmids, in conjunction with an improved helper plasmid, have improved the yield of recombinant AAV by up to 40-fold. Most likely a combination of strategies such as these will be necessary to develop a stable producer cell line that will be free of helper adenovirus and that will generate recombinant AAV at levels that are adequate for use in human gene therapies.

13.5 Direct Injection of Naked DNA

Direct injection of plasmid DNA intramuscularly, intraepidermally, and intravenously is a simple and direct technique for modifying cells *in vivo*. DNA injected intramuscularly or intraepidermally is internalized by cells proximal to the injection site [Hengge et al., 1995; Wolff et al., 1990]. DNA injected intravenously is rapidly degraded in the blood ($t_{1/2} < 10$ min.) or retained in various organs in the body, preferentially in the nonparenchymal cells of the liver [Kawabata et al., 1995; Lew et al., 1995]. Gene expression after direct injection has been demonstrated in skeletal muscle cells of rodents and nonhuman primates, cardiac muscle cells of rodents, livers of cats and rats, and in thyroid follicular cells of rabbits [Hengge et al., 1995; Hickman et al., 1994; Levine and Friedmann, 1993; Sikes et al., 1994].

The efficiency of gene transfer by direct injection is somewhat inefficient [Doh et al., 1997]. As little as 60 to 100 myocardial cells have been reported to be modified per injection [Acsadi et al., 1991]. Higher efficiencies were observed when plasmids were injected into regenerating muscle cells [Wells, 1993] or co-injected with recombinant adenoviruses [Yoshimura et al., 1993]. Injected DNA does not integrate, yet gene expression can persist for as long as two months [Wolff et al., 1990]. Levels of gene expression are often low but can be increased by improvements in vector design [Hartikka et al., 1996].

Despite the low efficiency of gene transfer and expression, direct injection of plasmid DNA has many promising applications in gene therapy, particularly when low levels of expression are sufficient to achieve the desired biological effect. For example, several patients suffering from critical limb ischemia were successfully treated by injection of DNA encoding human vascular endothelial growth factor (VEGF), a potent angiogenic factor [Isner et al., 1996; Isner et al., 1996; Melillo et al., 1997; Tsurumi et al., 1996]. Following injection of DNA into the ischemic limbs of ten patients, VEGF expression was detected in their serum, new blood vessels were formed in 7 of the 10 patients, and 3 patients were able to avoid scheduled below-the-knee amputations [Baumgartner et al., 1998].

Direct injection of DNA could also be used to vaccinate patients against pathogens as evidenced by the effects of direct injection of plasmid DNA encoding pathogen proteins or immunomodulatory cytokines [Donnelly et al., 1997; Sato et al., 1996]. Direct injection may also be an effective way to systemically deliver therapeutic proteins [Tripathy et al., 1996]. For example, the incidence of autoimmune diabetes in a mouse model of the disease was significantly reduced in mice that were injected intramuscularly with DNA encoding interleukin 10, an immunosuppressive cytokine [Nitta et al., 1998].

In contrast to viral-based delivery systems, there is little restriction on the size of the transgene that can be delivered by direct DNA injection. As a result, direct DNA injection is particularly well-suited for treating disorders that require the delivery of a large transgene. For example, Duchenne's muscular dystrophy, a genetic disease of the muscle caused by a defect in the gene for dystrophin (12 kilobases) can potentially be treated by direct DNA injection [Acsadi et al., 1991].

13.6 Particle-Mediated Gene Transfer

Particle-mediated gene transfer is an alternative method used to deliver plasmid DNA to cells. DNA coated gold particles are loaded onto a macro-projectile, which is then accelerated through a vacuum chamber to high velocity by a burst of helium or a voltage discharge until it hits a stopping plate [Yang et al., 1996]. Upon impact, the DNA-coated microprojectiles are released through a hole in the stopping plate, penetrate the target tissue, and the transferred gene is expressed [Morgan and Yarmush, 1998; Yang et al., 1996]. Genes have been introduced and expressed in a number of cell types and tissues, including skin, liver, spleen, muscle, intestine, hematopoietic cells, brain, oral mucosa and epidermis, tumor explants, and cells of developing mouse embryos [Jiao et al., 1993; Keller et al., 1996; Mahvi et al., 1996; Morgan and Yarmush, 1998; Verma et al., 1998; Zelenin et al., 1993]. Similar to direct DNA injection, there are few constraints on the size of the DNA that can be delivered.

The efficiency of particle-mediated gene transfer varies with tissue type, but in general it is most efficient in the liver, pancreas, and epidermis of the skin, and least efficient in muscle, vascular, and cardiac tissues [Rakhmilevich and Yang, 1997]. For example, 20 percent of bombarded epidermal cells of the skin, but only 1 to 3 percent of bombarded muscle cells, expressed the transferred gene [Williams et al., 1991; Yang et al., 1990].

13.7 Liposome-Mediated Gene Delivery

Liposomes made from a mixture of neutral and cationic lipids have also been used to deliver plasmid DNA to cells. Liposomes are relatively easy to make, can be made with well-defined biophysical properties, and can accommodate virtually any size transgene [Kay et al., 1997; Tomlinson and Rolland, 1996]. Small unilamellar liposomes ranging from 20 to 100 nm in diameter are prepared by sonication of a mixture of cationic (e.g., DOTMA, N-{1-(2,3-dioleyloxy)propyl}-N,N,N-triethylammonium) and neutral (e.g., DOPE, dioleoyl-phosphatidylethanolamine) lipids, followed by extrusion through a porous polycarbonate filter (e.g., 100 nm pore size) (Fig. 13.5) [Morgan and Yarmush, 1998; Radler

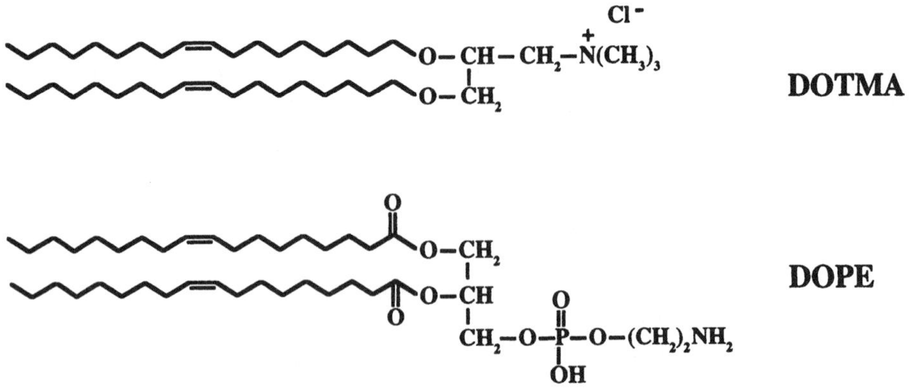

FIGURE 13.5 Common constituents of cationic liposomes. A typical cationic liposome is composed of a mixture of cationic lipids and neutral phospholipids. DOTMA {N-[1-(2,3,-dioleyloxy]-N,N,N-trimethylammonium chloride} is a synthetic cationic lipid that attaches to DNA using its positively charged head group. DOPE {1,2-Di[(*cis*)-9-octadecenoyl]-sn-glycero-3-phosphoethanolamine} is a neutral phospholipid which enhances the activity of these cationic liposomes. The resultant DNA-liposome complex is positively charged and reduces or eliminates the repulsive electrostatic forces between the negatively charged DNA and the negatively charged cell surface.

et al., 1997]. DNA is added to the cationic liposomes, binds noncovalently to the positively charged cationic lipids, and induces a topological transition from liposomes to multilamellar structures composed of lipid bilayers alternating with DNA monolayers [Felgner and Ringold, 1989; Radler et al., 1997]. The size of the structures depends on their overall charge, with charged structures (negative or positive) being the smallest (about 100 nm in diameter) due to stabilization by electrostatic repulsion and neutral structures being the largest (greater than 3000 nm in diameter) due to aggregation as a result of van der Waals attractive forces [Radler et al., 1997]. The relationship between structure and transfection efficiency is not well-understood, although in general a slight excess of cationic lipid is needed for optimal gene transfer [Kay et al., 1997].

Gene transfer is accomplished by simply mixing or applying the lipid-DNA complexes to the target cells or tissue. Cationic liposome-DNA complexes have been used in a number of applications, including transfer of genes to the arterial wall, lung, skin, and systemically by intravenous injection [Morgan and Yarmush, 1998; Zhu et al., 1993]. Genes delivered by cationic liposomes do not integrate, so gene expression is transient and there is minimal risk of insertional mutagenesis. Liposomes do not carry any viral sequences or proteins, and are relatively nontoxic and nonimmunogenic [Nabel et al., 1993; Zhu et al., 1993].

Liposome-mediated gene transfer is somewhat inefficient, however, in part due to the failure of a large fraction of lipid-complexed DNA to escape degradation in cellular endosomes [Xu and Szoka, 1996; Zabner et al., 1995]. Several strategies have been taken to overcome this limitation, including use of acidotropic bases to reduce the rate of degradation by raising the pH of the endosomes, coupling of liposomes to endosomolytic or fusogenic virus proteins, and developing new liposome formulations that use pH sensitive cationic lipids that become fusogenic in the acidic cellular endosomes [Budker et al., 1996; Legendre and Szoka, 1992; Yonemitsu et al., 1997]. Substantial improvements in the efficiency of liposome-mediated gene transfer will only be achieved when the fundamental mechanisms that govern the interactions of cationic lipid-DNA complexes with cells are better understood.

13.8 Other Gene Transfer Methods

Several other gene transfer technologies have been tested in clinical trials or are in various stages of development. These include recombinant viruses such as vaccinia virus [Qin and Chatterjee, 1996], herpes simplex virus [Glorioso et al., 1997], canarypox virus [Kawakita et al., 1997], and fowlpox [Wang et al., 1995], and nonviral vectors that mimic some of the properties of viruses, such as DNA conjugated to proteins that promote binding to specific cell-surface receptors, fusion, or localization to the nucleus [Wu et al., 1994]. These methods are not discussed here in greater detail because even though they are capable of transferring genes to cells they have not yet been well-developed or extensively tested in the clinic.

13.9 Summary and Conclusion

Several gene transfer systems have been developed that have successfully transferred genes to cells, and have elicited various biological effects. To date, nearly 300 clinical trials have been approved to test their safety and efficacy (Table 13.4). Each system has unique features, advantages, and disadvantages that determine if its use in a particular application is appropriate. One principal consideration is whether or not permanent or temporary genetic modification is desired. Other important considerations include what the setting of gene transfer will be (*ex vivo* or *in vivo*), what level of gene expression is needed, and whether or not the host has preexisting immunity against the vector. No single gene transfer system is ideal for any particular application and it is unlikely that such a universal gene transfer system will ever be developed. More likely, the current gene transfer systems will be further improved and modified, and new systems developed, that are optimal for the treatment of specific diseases.

TABLE 13.4   Current Clinical Experience Using Gene Transfer (as of December 1, 1998)

Human gene transfer protocols	Number	Percentage
Gene marking studies	30	11
Nontherapeutic protocols	2	1
Therapeutic protocols	238	88
Infectious diseases	25	9
Inherited diseases	35	13
Cancer	167	62
Other diseases	11	4
Strategies for gene delivery		
Ex vivo protocols	134	50
In vivo protocols	136	50
Gene transfer technologies		
Permanent genetic modification		
Recombinant retroviruses	128	47
Recombinant adeno-associated virus	6	2
Temporary genetic modification		
Recombinant adenovirus	60	22
Cationic liposomes	37	14
Plasmid DNA	16	6
Particle mediated	2	1
Other		
Viral vectors	19	7
Nonviral vector	2	1

*"Other" viral vectors refers to recombinant canarypox virus, vaccinia virus, fowlpox, and herpes simplex virus. "Other" nonviral vector refers to RNA transfer and antisense delivery.

Defining Terms

Chimeric retrovirus: A recombinant retrovirus whose structural proteins are derived from two or more different viruses.

Ex vivo: Outside the living body, referring to a process or reaction occurring therein.

In vitro: In an artificial environment, referring to a process or reaction occurring therein, as in a test tube or culture dish.

In vivo: In the living body, referring to a process or reaction occurring therein.

Liposome: A spherical particle of lipid substance suspended in an aqueous medium.

Packaging cell line: Cells that express all the structural proteins required to form an infectious viral particle.

Plasmid: A small, circular extrachromosomal DNA molecule capable of independent replication in a host cell, typically a bacterial cell.

Recombinant: A virus or vector that has DNA sequences not originally (naturally) present in their DNA.

Retrovirus: A virus that possesses RNA-dependent DNA polymerase (reverse transcriptase) that reverse transcribes the virus' RNA genome into DNA, then integrates that DNA into the host cell's genome .

Transduce: To effect transfer and integration of genetic material to a cell by infection with a recombinant retrovirus.

Vector: A plasmid or viral DNA molecule into which a DNA sequence (typically encoding a therapeutic protein) is inserted.

References

Acsadi G, Dickson G, Love DR et al., 1991. Human dystrophin expression in mdx mice after intramuscular injection of DNA constructs [see comments]. *Nature.* 352:815-8.

Acsadi G, Jiao SS, Jani A et al., 1991. Direct gene transfer and expression into rat heart *in vivo*. *New Biol.* 3:71-81.

Anderson WF. 1992. Human gene therapy. *Science*. 256:808-13.

Andreadis ST, Roth CM, Le Doux JM et al., 1999. Large scale processing of recombinant retroviruses for gene therapy. *Biotechnol. Prog.* Jan-Feb; 15(1):1–11.

Bartlett JS, Samulski RJ, and McCown TJ. 1998. Selective and rapid uptake of adeno-associated virus type 2 in brain. *Hum. Gene Ther.* 9:1181-6.

Baumgartner I, Pieczek A, Manor O et al., 1998. Constitutive expression of phVEGF165 after intramuscular gene transfer promotes collateral vessel development in patients with critical limb ischemia [see comments]. *Circulation*. 97:1114-23.

Boucher RC, and Knowles MR. 1993. Gene therapy for cystic fibrosis using E1 deleted adenovirus: A phase I trial in the nasal cavity. Office of Recombinant DNA Activity, NIH.

Budker V, Gurevich V, Hagstrom JE et al., 1996. pH-sensitive, cationic liposomes: a new synthetic virus-like vector. *Nat. Biotechnol.* 14:760-4.

Burns JC, Friedmann T, Driever W et al., 1993. Vesicular stomatitis virus G glycoprotein pseudotyped retroviral vectors: concentration to very high titer and efficient gene transfer into mammalian and nonmammalian cells [see comments]. *Proc. Natl. Acad. Sci. USA*. 90:8033-7.

Capecchi MR. 1980. High efficiency transformation by direct microinjection of DNA into cultured mammalian cells. *Cell*. 22:479-88.

Chen C, and Okayama H. 1987. High-efficiency transformation of mammalian cells by plasmid DNA. *Mol. Cell Biol.* 7:2745-52.

Crystal RG. 1995. Transfer of genes to humans: early lessons and obstacles to success. *Science*. 270:404-10.

Dai Y, Schwarz EM, Gu D et al., 1995. Cellular and humoral immune responses to adenoviral vectors containing factor IX gene: tolerization of factor IX and vector antigens allows for long-term expression. *Proc. Natl. Acad. Sci. USA*. 92:1401-5.

Danos O, and Mulligan RC. 1988. Safe and efficient generation of recombinant retroviruses with amphotropic and ecotropic host ranges. *Proc. Natl. Acad. Sci. USA*. 85:6460-4.

Doh SG, Vahlsing HL, Hartikka J et al., 1997. Spatial-temporal patterns of gene expression in mouse skeletal muscle after injection of lacZ plasmid DNA. *Gene Ther.* 4:648-63.

Donnelly JJ, Ulmer JB, Shiver JW et al., 1997. DNA vaccines. *Ann. Rev. Immunol.* 15:617-48.

Engelhardt JF, Ye X, Doranz B et al., 1994. Ablation of E2A in recombinant adenoviruses improves transgene persistence and decreases inflammatory response in mouse liver. *Proc. Natl. Acad. Sci. USA*. 91:6196-200.

Felgner PL, and Rhodes G. 1991. Gene therapeutics. *Nature*. 349:351-2.

Felgner PL, and Ringold GM. 1989. Cationic liposome-mediated transfection. *Nature*. 337:387-8.

Ferrari FK, Xiao X, McCarty D et al., 1997. New developments in the generation of Ad-free, high-titer rAAV gene therapy vectors. *Nat. Med.* 3:1295-7.

Flannery JG, Zolotukhin S, Vaquero MI et al., 1997. Efficient photoreceptor-targeted gene expression *in vivo* by recombinant adeno-associated virus. *Proc. Natl. Acad. Sci. USA*. 94:6916-21.

Flotte TR, Afione SA, Conrad C et al., 1993. Stable *in vivo* expression of the cystic fibrosis transmembrane conductance regulator with an adeno-associated virus vector. *Proc. Natl. Acad. Sci. USA*. 90:10613-7.

Friedmann T. 1989. Progress toward human gene therapy. *Science*. 244:1275-81.

Glorioso JC, Goins WF, Schmidt MC et al., 1997. Engineering herpes simplex virus vectors for human gene therapy. *Adv. Pharmacol.* 40:103-36.

Greber UF, Willetts M, Webster P et al., 1993. Stepwise dismantling of adenovirus 2 during entry into cells. *Cell*. 75:477-86.

Hartikka J, Sawdey M, Cornefert-Jensen F et al., 1996. An improved plasmid DNA expression vector for direct injection into skeletal muscle. *Hum. Gene. Ther.* 7:1205-17.

Hengge UR, Chan EF, Foster RA et al., 1995. Cytokine gene expression in epidermis with biological effects following injection of naked DNA. *Nat. Genet.* 10:161-6.

Hickman MA, Malone RW, Lehmann-Bruinsma K et al., 1994. Gene expression following direct injection of DNA into liver. *Hum. Gene. Ther.* 5:1477-83.

Hitt MM, Addison CL, and Graham FL. 1997. Human adenovirus vectors for gene transfer into mammalian cells. *Adv. Pharmacol.* 40:137-206.

Horwitz MS. 1990. Adenoviridae and their replication, p. 1679. In BN Fields, et al. (ed.), *Fields Virology*, 2nd ed, vol. 2. Raven Press, Ltd., New York.

Inoue N, and Russell DW. 1998. Packaging cells based on inducible gene amplification for the production of adeno-associated virus vectors. *J. Virol.* 72:7024-7031.

Isner JM, Pieczek A, Schainfeld R et al., 1996. Clinical evidence of angiogenesis after arterial gene transfer of phVEGF165 in patient with ischaemic limb [see comments]. *Lancet.* 348:370-4.

Isner JM, Walsh K, Symes J et al., 1996. Arterial gene transfer for therapeutic angiogenesis in patients with peripheral artery disease. *Hum. Gene. Ther.* 7:959-88.

Jiao S, Cheng L, Wolff JA et al., 1993. Particle bombardment-mediated gene transfer and expression in rat brain tissues. *Biotechnology* (NY). 11:497-502.

Kawabata K, Takakura Y, and Hashida M. 1995. The fate of plasmid DNA after intravenous injection in mice: involvement of scavenger receptors in its hepatic uptake. *Pharm. Res.* 12:825-30.

Kawakita M, Rao GS, Ritchey JK et al., 1997. Effect of canarypox virus (ALVAC)-mediated cytokine expression on murine prostate tumor growth [see comments]. *J. Natl. Cancer Inst.* 89:428-36.

Kay MA, Holterman AX, Meuse L et al., 1995. Long-term hepatic adenovirus-mediated gene expression in mice following CTLA4Ig administration. *Nat. Genet.* 11:191-7.

Kay MA, Liu D, and Hoogerbrugge PM. 1997. Gene therapy. *Proc. Natl. Acad. Sci. USA.* 94:12744-6.

Keller ET, Burkholder JK, Shi F et al., 1996. *In vivo* particle-mediated cytokine gene transfer into canine oral mucosa and epidermis. *Cancer Gene Ther.* 3:186-91.

Kotani H, Newton PB, 3rd, Zhang S et al., 1994. Improved methods of retroviral vector transduction and production for gene therapy. *Hum. Gene. Ther.* 5:19-28.

Le Doux JM, Davis HE, Morgan JR et al., 1999. Kinetics of retrovirus production and decay. *Biotech. Bioeng. Prog.* In Press.

Le Doux JM, Morgan JR, and Yarmush ML. 1998. Removal of proteoglycans increases efficiency of retroviral gene transfer. *Biotechnol. Bioeng.* 58:23-34.

Ledley FD. 1993. Hepatic gene therapy: present and future. *Hepatology.* 18:1263-73.

Legendre JY, and Szoka FC, Jr. 1992. Delivery of plasmid DNA into mammalian cell lines using pH-sensitive liposomes: comparison with cationic liposomes. *Pharm. Res.* 9:1235-42.

Levine F, and Friedmann T. 1993. Gene therapy. *Am. J. Dis. Child.* 147:1167-74.

Lew D, Parker SE, Latimer T et al., 1995. Cancer gene therapy using plasmid DNA: Pharmacokinetic study of DNA following injection in mice [see comments]. *Hum. Gene. Ther.* 6:553-64.

Maeda Y, Ikeda U, Shimpo M et al., 1998. Efficient gene transfer into cardiac myocytes using adeno-associated virus (AAV) vectors [In Process Citation]. *J. Mol. Cell Cardiol.* 30:1341-8.

Magovern CJ, Mack CA, Zhang J et al., 1997. Regional angiogenesis induced in nonischemic tissue by an adenoviral vector expressing vascular endothelial growth factor. *Hum. Gene. Ther.* 8:215-27.

Mahvi DM, Burkholder JK, Turner J et al., 1996. Particle-mediated gene transfer of granulocyte-macrophage colony- stimulating factor cDNA to tumor cells: implications for a clinically relevant tumor vaccine. *Hum. Gene. Ther.* 7:1535-43.

McCarty DM, and Samulski RJ. 1997. Adeno-associated viral vectors. In M Strauss, et al. (eds.), *Concepts in Gene Therapy*, p. 61-78. Berlin, Walter de Gruyter & Co.

McGrath M, Witte O, Pincus T et al., 1978. Retrovirus purification: Method that conserves envelope glycoprotein and maximizes infectivity. *J. Virol.* 25:923-7.

Melillo G, Scoccianti M, Kovesdi I et al., 1997. Gene therapy for collateral vessel development. *Cardiovasc. Res.* 35:480-9.

Miller AD. 1992. Retroviral vectors. *Curr. Top. Microbiol. Immunol.* 158:1-24.

Monahan PE, Samulski RJ, Tazelaar J et al., 1998. Direct intramuscular injection with recombinant AAV vectors results in sustained expression in a dog model of hemophilia. *Gene Ther.* 5:40-9.

Morgan JR, Tompkins RG, and Yarmush ML. 1993. Advances in recombinant retroviruses for gene delivery. *Adv. Drug Delivery Rev.* 12:143-158.

Morgan JR, Tompkins RG, and Yarmush ML. 1994. Genetic engineering and therapeutics. In RS Greco (ed.), *Implantation Biology: The Host Responses and Biomedical Devices,* p. 387. CRC Press, Boca Raton, FL.

Morgan JR, and Yarmush ML. 1998. Gene therapy in tissue engineering, In CW Patrick Jr., et al. (eds.), *Frontiers in Tissue Engineering.* p. 278-310. Elsevier Science Ltd, Oxford.

Morsy MA, Gu M, Motzel S et al., 1998. An adenoviral vector deleted for all viral coding sequences results in enhanced safety and extended expression of a leptin transgene. *Proc. Natl. Acad. Sci. USA.* 95:7866-71.

Mulligan RC. 1993. The basic science of gene therapy. *Science.* 260:926-32.

Murphy JE, Zhou S, Giese K et al., 1997. Long-term correction of obesity and diabetes in genetically obese mice by a single intramuscular injection of recombinant adeno-associated virus encoding mouse leptin. *Proc. Natl. Acad. Sci. USA.* 94:13921-6.

Muzyczka N. 1992. Use of adeno-associated virus as a general transduction vector for mammalian cells. *Curr. Top. Microbiol. Immunol.* 158:97-129.

Nabel GJ, Nabel EG, Yang ZY et al., 1993. Direct gene transfer with DNA-liposome complexes in melanoma: expression, biologic activity, and lack of toxicity in humans. *Proc. Natl. Acad. Sci. USA.* 90:11307-11.

Nakagawa I, Murakami M, Ijima K et al., 1998. Persistent and secondary adenovirus-mediated hepatic gene expression using adenovirus vector containing CTLA4IgG [In Process Citation]. *Hum. Gene. Ther.* 9:1739-45.

Naldini L, Blomer U, Gallay P et al., 1996. *In vivo* gene delivery and stable transduction of nondividing cells by a lentiviral vector [see comments]. *Science.* 272:263-7.

Neumann E, Schaefer-Ridder M, Wang Y et al., 1982. Gene transfer into mouse lyoma cells by electroporation in high electric fields. *Embo J.* 1:841-5.

Nitta Y, Tashiro F, Tokui M et al., 1998. Systemic delivery of interleukin 10 by intramuscular injection of expression plasmid DNA prevents autoimmune diabetes in nonobese diabetic mice. *Hum. Gene. Ther.* 9:1701-7.

Pagano JS, McCutchan JH, and Vaheri A. 1967. Factors influencing the enhancement of the infectivity of poliovirus ribonucleic acid by diethylaminoethyl-dextran. *J. Virol.* 1:891-7.

Parks RJ, Chen L, Anton M et al., 1996. A helper-dependent adenovirus vector system: Removal of helper virus by Cre-mediated excision of the viral packaging signal. *Proc. Natl. Acad. Sci. USA.* 93:13565-70.

Paul RW, Morris D, Hess BW et al., 1993. Increased viral titer through concentration of viral harvests from retroviral packaging lines. *Hum. Gene. Ther.* 4:609-15.

Qin H, and Chatterjee SK. 1996. Cancer gene therapy using tumor cells infected with recombinant vaccinia virus expressing GM-CSF. *Hum. Gene. Ther.* 7:1853-60.

Radler JO, Koltover I, Salditt T et al., 1997. Structure of DNA-cationic liposome complexes: DNA intercalation in multilamellar membranes in distinct interhelical packing regimes [see comments]. *Science.* 275:810-4.

Rakhmilevich AL, and Yang N-S. 1997. Particle-mediated gene delivery system for cancer research. In M Strauss, et al. (eds.), *Concepts in Gene Therapy,* p. 109-120, Berlin, Walter de Gruyter.

Roe T, Reynolds TC, Yu G et al., 1993. Integration of murine leukemia virus DNA depends on mitosis. *Embo. J.* 12:2099-108.

Roemer K, and Friedmann T. 1992. Concepts and strategies for human gene therapy. *Eur. J. Biochem.* 208:211-25.

Sato Y, Roman M, Tighe H et al., 1996. Immunostimulatory DNA sequences necessary for effective intradermal gene immunization. *Science.* 273:352-4.

Scaria A, St George JA, Jiang C et al., 1998. Adenovirus-mediated persistent cystic fibrosis transmembrane conductance regulator expression in mouse airway epithelium. *J. Virol.* 72:7302-9.

Sikes ML, O. Malley BW J, Finegold MJ et al., 1994. *In vivo* gene transfer into rabbit thyroid follicular cells by direct DNA injection. *Hum. Gene. Ther.* 5:837-44.

Temin HM. 1990. Safety considerations in somatic gene therapy of human disease with retrovirus vectors. *Hum. Gene. Ther.* 1:111-23.

Tomlinson E, and Rolland AP. 1996. Controllable gene therapy pharmaceutics of non-viral gene delivery systems. *J. Contr. Rel.* 39:357-372.

Tripathy SK, Svensson EC, Black HB et al., 1996. Long-term expression of erythropoietin in the systemic circulation of mice after intramuscular injection of a plasmid DNA vector. *Proc. Natl. Acad. Sci. USA.* 93:10876-80.

Tsurumi Y, Takeshita S, Chen D et al., 1996. Direct intramuscular gene transfer of naked DNA encoding vascular endothelial growth factor augments collateral development and tissue perfusion [see comments]. *Circulation.* 94:3281-90.

Verma IM, and Somia N. 1997. Gene therapy — Promises, problems and prospects [news]. *Nature.* 389:239-42.

Verma S, Woffendin C, Bahner I et al., 1998. Gene transfer into human umbilical cord blood-derived CD34+ cells by particle-mediated gene transfer [In Process Citation]. *Gene Ther.* 5:692-9.

Wang M, Bronte V, Chen PW et al., 1995. Active immunotherapy of cancer with a nonreplicating recombinant fowlpox virus encoding a model tumor-associated antigen. *J. Immunol.* 154:4685-92.

Weiss R, Teich N, Varmus H et al., 1982. Molecular biology of tumor viruses. Cold Spring Harbor Laboratory, Cold Spring Harbor.

Wells DJ. 1993. Improved gene transfer by direct plasmid injection associated with regeneration in mouse skeletal muscle. *FEBS Lett.* 332:179-82.

Welsh MJ. 1993. Adenovirus-mediated gene transfer of CFTR to the nasal epithelium and maxillary sinus of patients with cystic fibrosis. Office of Recombinant DNA Activity, NIH.

Wickham TJ, Mathias P, Cheresh DA et al., 1993. Integrins alpha v beta 3 and alpha v beta 5 promote adenovirus internalization but not virus attachment. *Cell.* 73:309-19.

Williams RS, Johnston SA, Riedy M et al., 1991. Introduction of foreign genes into tissues of living mice by DNA-coated microprojectiles. *Proc. Natl. Acad. Sci. USA.* 88:2726-30.

Wilmott RW, and Whitsett J. 1993. A phase I study of gene therapy of cystic fibrosis utilizing a replication deficient recombinant adenovirus vector to deliver the human cystic fibrosis transmembrane conductance regulator cDNA to the airways. Office of Recombinant DNA Activity, NIH.

Wold WS, and Gooding LR. 1989. Adenovirus region E3 proteins that prevent cytolysis by cytotoxic T cells and tumor necrosis factor. *Mol. Biol. Med.* 6:433-52.

Wolff JA, Malone RW, Williams P et al., 1990. Direct gene transfer into mouse muscle *in vivo*. *Science.* 247:1465-8.

Wu GY, Zhan P, Sze LL et al., 1994. Incorporation of adenovirus into a ligand-based DNA carrier system results in retention of original receptor specificity and enhances targeted gene expression. *J. Biol. Chem.* 269:11542-6.

Xiao W, Berta SC, Lu MM et al., 1998. Adeno-associated virus as a vector for liver-directed gene therapy [In Process Citation]. *J. Virol.* 72:10222-6.

Xiao X, Li J, and Samulski RJ. 1998. Production of high-titer recombinant adeno-associated virus vectors in the absence of helper adenovirus. *J. Virol.* 72:2224-32.

Xu Y, and Szoka FC, Jr. 1996. Mechanism of DNA release from cationic liposome/DNA complexes used in cell transfection. *Biochem.* 35:5616-23.

Yang NS, Burkholder J, Roberts B et al., 1990. *In vivo* and *in vitro* gene transfer to mammalian somatic cells by particle bombardment. *Proc. Natl. Acad. Sci. USA.* 87:9568-72.

Yang NS, Sun WH, and McCabe D. 1996. Developing particle-mediated gene-transfer technology for research into gene therapy of cancer. *Mol. Med. Today.* 2:476-81.

Yonemitsu Y, Kaneda Y, Muraishi A et al., 1997. HVJ (Sendai virus)-cationic liposomes: a novel and potentially effective liposome-mediated technique for gene transfer to the airway epithelium. *Gene Ther.* 4:631-8.

York IA, Roop C, Andrews DW et al., 1994. A cytosolic herpes simplex virus protein inhibits antigen presentation to CD8+ T lymphocytes. *Cell.* 77:525-35.

Yoshimura K, Rosenfeld MA, Seth P et al., 1993. Adenovirus-mediated augmentation of cell transfection with unmodified plasmid vectors. *J. Biol. Chem.* 268:2300-3.

Zabner J, Fasbender AJ, Moninger T et al., 1995. Cellular and molecular barriers to gene transfer by a cationic lipid. *J. Biol. Chem.* 270:18997-9007.

Zelenin AV, Alimov AA, Zelenina IA et al., 1993. Transfer of foreign DNA into the cells of developing mouse embryos by microprojectile bombardment. *FEBS Lett.* 315:29-32.

Zhu N, Liggitt D, Liu Y et al., 1993. Systemic gene expression after intravenous DNA delivery into adult mice. *Science.* 261:209-11.

Zsengeller ZK, Wert SE, Hull WM et al., 1995. Persistence of replication-deficient adenovirus-mediated gene transfer in lungs of immune-deficient (nu/nu) mice. *Hum. Gene. Ther.* 6:457-67.

14

Cell Engineering

Douglas A. Lauffenburger
*Massachusetts Institute
of Technology*

Cell engineering can be defined either academically, as "application of the principles and methods of engineering to the problems of cell and molecular biology of both a basic and applied nature" [Nerem, 1991], or functionally, as "manipulation of cell function by molecular approaches" [Lauffenburger and Aebischer, 1993]. However, defined, there is no question that cell engineering is becoming a central area of biomedical engineering. Applications to health care technology include design of molecular therapies for wound healing, cancer, and inflammatory disease, development of biomaterials and devices for tissue regeneration and reconstruction, introduction of gene therapies to remedy a variety of disorders, and utilization of mammalian cell culture technology for production of therapeutic proteins.

Cell engineering has its roots in revolutions that have occurred in the field of molecular cell biology over the past few decades (see Darnell et al., [1986, Chap. 1] for an excellent historical summary). Key advances upon which this field was built include electron microscopy in the 1940s, permitting intracellular structural features to be seen; the identification of DNA as the biochemical basis of genetics, also in the 1940s; the elucidation of the structure and function of DNA in the 1950s; the discovery of gene regulation in the 1960s; and progress in areas such as molecular genetics, light microscopy, and protein biochemistry in the 1970s, 1980s, and 1990s. Combining the powerful set of information and techniques from molecular cell biology with the analytical and synthetic approaches of engineering for purposeful exploitation and manipulation of cell function has given rise to the wide range of applications listed above. Engineering contributions include elucidation of kinetic, transport, and mechanical effects; identification and measurement of system parameters; and production of novel materials and devices required for control of cell and tissue functions by molecular mechanisms. All these aspects of engineering are essential to development of reliable technological systems requiring reproducible performance of cells and tissues.

The central paradigm of cell engineering is employment of an engineering perspective of understanding system function in terms of underlying component properties, with the cells and tissues being the system and the molecular mechanisms being the components. Thus cell and tissue functions must be quantified and related to molecular properties that serve as the key design parameters. This paradigm is made especially powerful by the capability derived from molecular cell biology to alter molecular properties intentionally, permitting rational system design based on the identified parameters.

14.1 Basic Principles

The fundamental aspects of cell engineering are twofold: (1) quantitative understanding of cell function in molecular terms and (2) ability to manipulate cell function through molecular mechanisms, whether by pharmacologic or genetic intervention or by introduction of materials or devices. Elucidation of basic technical principles that will stand a relatively long-term test of time is difficult because the field of molecular cell biology continues to experience incredibly rapid and unpredictable advances. However, an attempt can be made to at least outline central principles along the two fundamental lines cited above. First, one can consider the basic categories of cell function that may be encountered in cell engineering applications. These include proliferation, adhesion, migration, uptake and secretion, and differentiation. Second, one can consider basic categories of regulation of these functions. Two major loci of regulation are gene expression and structural/enzymatic protein activity; these processes act over relatively long-term (hours) and short-term (seconds to minutes) time spans, respectively. Thus a cell can change the identities and quantities of the molecular components it makes as well as their functional productivity. A third locus of regulation, which essentially encompasses the others, is receptor-mediated signaling. Molecular ligands interacting with cell receptors generate intracellular chemical and mechanical signals that govern both gene expression and protein activity and thus essentially control cell function.

At the present time, bioengineering efforts to manipulate cell behavior in health care technologies are primarily centered on development of materials, devices, and molecular or cellular therapies based on receptor-ligand interactions. These interactions are amenable to engineering approaches when relevant biochemical and biophysical properties of receptors and their ligands are identified [Lauffenburger and Linderman, 1993]. Examples of such efforts and the basic principles known at the present time to be involved are described below in three categories of current major activity: cell proliferation, cell adhesion, and cell migration.

14.2 Cell Proliferation

Proliferation of mammalian tissue and blood cells is regulated by signaling of growth factor receptors following ligand binding. In some cases, the resulting regulation is that of an on/off switch, in which the fraction of cells in a population that are stimulated to move into a proliferative state is related to the level of receptor occupancy; an example of this is the control of blood cell proliferation in the bone marrow [Kelley et al., 1993]. In other cases, the rate of progression through the cell cycle, governed by the rate of movement into the DNA synthesis phase, is related to the level of receptor occupancy. For two examples at least, epidermal growth factor stimulation of fibroblast proliferation [Knauer et al., 1984] and interleukin 2 stimulation of T-lymphocyte proliferation [Robb, 1982], DNA synthesis is linearly proportional to the number of growth factor-receptor complexes, perhaps with a minimum threshold level.

In engineering terms, two key principles emerge from this relationship. One is that the details of the biochemical signal transduction cascade are reflected in the coefficient relating DNA synthesis to the number of complexes, which represents an intrinsic mitogenic sensitivity. Thus, for a specified number of complexes, the cell proliferation rate can be altered by modifying aspects of the signal transduction cascade. The second principle is that mechanisms governing the number, and perhaps location, of growth factor-receptor complexes can influence cell proliferation with just as great an effect as signal transduction mechanisms. At the present time, it is more straightforward for cell engineering purposes to manipulate cell proliferation rates by controlling the number of signaling complexes using a variety of methods. Attempts to quantitatively analyze and manipulate components of receptor signal transduction cascades are also beginning [Mahama and Linderman, 1994; Renner et al., 1994].

Engineering methods helpful for providing desired growth factor concentrations include design of controlled-release devices [Powell et al., 1990] and biomaterials [Cima, 1994]. Endogenous alteration of ligand concentration and receptor number can be accomplished using gene transfer methods to introduce DNA that can increase ligand production or receptor expression or antisense therapy to introduce RNA that can decrease ligand production or reduce receptor expression. Levels of growth factor-receptor binding can be tuned by rational design of ligands; these might be antagonists that block binding of the normal growth factor and hence decrease complex levels or super-agonist growth factor mimics that provide for increased levels of complex formation over that of the normal growth factor. Rates or extents of receptor and ligand trafficking processes can be altered by modification of the receptor, ligand, or accessory components. Such alterations can affect the number of signaling complexes by increasing or decreasing the amount of receptor downregulation, ligand depletion, or compartmentation of complexes from the cell surface to intracellular locations.

Regulation of fibroblast proliferation by epidermal growth factor (EGF) offers an instructive example of some of these approaches. Under normal circumstances, fibroblasts internalize EGF-receptor complexes efficiently via ligand-induced endocytosis, leading to substantial downregulation of the EGF-receptor and depletion of EGF from the extracellular medium. The latter phenomenon helps account for the requirement for periodic replenishment of serum in cell culture; although growth factors present in serum could in principle operate as catalysts generating mitogenic signals through receptor binding, they generally behave as pseudonutrients because of endocytic internalization and subsequent degradation. Use of a controlled-release device for EGF delivery thus improved wound-healing responses compared with topical administration by helping compensate for growth factor depletion due to cell degradation as well as proteolysis in the tissue and loss into the bloodstream [Buckley et al., 1985]. Transfecting the gene for EGF into fibroblasts can allow them to become a self-stimulating autocrine system not requiring exogenous supplementation [Will et al., 1995]. Altering EGF/EGF-receptor trafficking can be similarly useful. Fibroblasts possessing variant EGF receptors for which ligand-induced endocytic internalization is abrogated exhibit dramatically increased proliferation rates at low EGF concentrations, primarily due to a diminished depletion of EGF from the medium [Reddy et al., 1994]. Increasing the proportion of ligands or receptors sorted in the endosome to recycling instead of degradation also reduces ligand depletion and receptor downregulation, thus increasing the number of growth factor-receptor complexes. An instance of this is the difference in response of cells to transforming growth factor alpha compared with EGF [French et al., 1994]; the receptor-binding affinity of TGFα is diminished at endosomal pH, permitting most of the TGFα-receptor complexes to dissociate with consequently increased receptor recycling, in contrast to EGF, which remains predominantly bound to the receptor in the endosome, leading to receptor degradation. TGFα has been found to be a more potent wound-healing agent, likely due to this alteration in trafficking [Schultz et al., 1987]. Hence ligands can be designed to optimize trafficking processes for most effective use in wound-healing, cell culture, and tissue engineering applications.

Proliferation of many tissue cell types is additionally regulated by attachment to extracellular matrices via interactions between adhesion receptors and their ligands. The quantitative dependence of proliferation on adhesion receptor-ligand interactions is not understood nearly as well as that for growth factors, though a relationship between the extent of cell spreading and DNA synthesis has been elucidated [Ingber et al., 1987] and has been applied in qualitative fashion for engineering purposes [Singhvi et al., 1994]. It is likely that a quantitative dependence of proliferation on adhesion receptor-ligand binding, or a related quantity, analogous to that found for growth factors will soon be developed and exploited in similar fashion by design of attachment substrata with appropriate composition of attachment factors. Properties of the cell environment affecting molecular transport [Yarmush et al., 1992] and mechanical stresses [Buschmann et al., 1992] are also important in controlling cell proliferation responses.

14.3 Cell Adhesion

Attachment of mammalian tissue and blood cells to other cells, extracellular matrices, and biomaterials is controlled by members of the various families of adhesion receptors, including the integrins, selectins, cadherins, and immunoglobulins. Interaction of these receptors with their ligands — which are typically accessed in an insoluble context instead of free in solution — is responsible for both physical attachment and chemical signaling events. (For clarity of discussion, the term *receptor* will refer to the molecules on "free" cells, while the term *ligand* will refer to the complementary molecules present on a more structured substratum, even if that substratum is a cell layer such as blood vessel endothelium.) It is essential to note at the outset that biologic adhesion is a dissipative phenomenon, one that in general is nonreversible; the energy of adhesion is typically less than the energy of deadhesion. Thus cell adhesion cannot be analyzed solely in terms of a colloidal adhesion framework.

Cell spreading following adhesion to a substratum is an active process requiring expenditure of energy for membrane protrusion and intracellular (cytoskeletal) motile force generation rather than being a purely passive, thermodynamic phenomenon. Similarly, migration of cells over two-dimensional and through three-dimensional environments requires involvement of adhesion receptors not only for physical transmission of motile force via traction to the substratum but also perhaps for active signaling leading to generation of the cytoskeletal motile force.

An engineering approach can be applied to design of cells, ligands, and ligand-bearing biomaterials to optimize attachment and subsequent signaling. For instance, material bearing synthetic peptides to which only a specific, desired cell type will adhere can be developed; an example is the immobilization of a unique amino acid sequence on a vascular-graft polymer to promote attachment and spreading of endothelial cells but not fibroblasts, smooth muscle cells, or platelets [Hubbell et al., 1991]. Moreover, the substratum density of a particular adhesive ligand required for proper cell spreading and leading to desired behavior can be quantitatively determined [Massia and Hubbell, 1991]. Design of substratum matrices with appropriate ligands for attachment, spreading, and proliferation in wound-healing and tissue-generation applications can be based on such information. These matrices also must possess suitable biodegradation properties along with appropriate mechanical properties for coordination with forces generated by the cells themselves as well as external loads [Cima et al., 1991b, Tranquillo et al., 1992].

Analyses of cell-substratum adhesion phenomena have elucidated key design parameters of receptor-ligand interactions as well as cell and substratum mechanics. An important principle is that the strength of adhesion is proportional to the receptor-ligand equilibrium binding affinity in a logarithmic manner [Kuo and Lauffenburger, 1993], so to alter adhesion strength by an order of magnitude requires a change in binding affinity by a factor of roughly 10^3. Cells possess the capability to accomplish this via covalent modification of proteins, leading to changes in effective affinity (i.e., avidity) either through direct effects on receptor-ligand binding or through indirect effects in which the state of receptor aggregation is modified. Aggregation of adhesion receptors into organized structures termed *focal contacts* leads to dramatic increases in adhesion strength because distractive forces are distributed over many receptor-ligand bonds together, as demonstrated experimentally [Lotz et al., 1989] and explained theoretically [Ward and Hammer, 1993]. Hence biochemical processes can lead to changes in biophysical properties. The fact that physical forces can affect distribution of receptor-ligand bonds and probably other protein-protein interactions suggests that biophysical processes can similarly alter cellular biochemistry [Ingber, 1993].

Cells are able to regulate the mode of detachment from a substratum, with a variety of alternative sites at which the linkage can be disrupted. Between the intracellular cytoskeleton and the extracellular matrix are generally a number of intermolecular connections involving the adhesion receptor. The receptor can be simply associated with membrane lipids or can be anchored by linkages to one or more types of intracellular components, some of which are also linked to actin filaments. Thus cell-substratum detachment may take place by simple ligand-receptor dissociation or extraction of the intact ligand-receptor bond from the membrane or extraction of the bond along with some intracellularly linked components. Biochemical regulation by the cell itself, likely involving covalent modification of the receptor and associated components, may serve as a means to control the mode of detachment when that is an important issue for engineering purposes. It is likely that the different superfamilies of adhesion receptors possess distinct characteristics of their micromechanical and biochemical regulation properties leading to different responses to imposed stresses generated either intracellularly or extracellularly.

A different situation involving the cell adhesion is that of cell capture by a surface from a flowing fluid. Instances of this include homing of white blood cells to appropriate tissues in the inflammatory and immune responses and metastatic spread of tumor cells, both by adhesion to microvascular endothelial cells, as well as isolation of particular cell subpopulations by differential cell adhesiveness to ligand-coated materials to facilitate cell-specific therapies. In such cases, it may be desired to express appropriate attachment ligands at required densities on either the target cells or capturing surfaces or to use soluble receptor-binding competitors to block attachment in the circulation. This is a complicated problem involving application of quantitative engineering analysis of receptor-ligand binding kinetics as well as cell and molecular mechanics. Indeed, properties of adhesion receptors crucial to facilitating attachment from flowing fluid are still uncertain; they appear to include the bond-formation rate constant but also bond mechanical parameters such as strength (i.e., ability to resist dissociation or distraction from the cell membrane) and compliance (i.e., ability to transfer stress into bond-dissociating energy).

Cell-substratum attachment during encounter in fluid flow conditions can be considered under equilibrium control if the strength or affinity of bonds is the controlling parameter, whereas it is under kinetic control when the rate of bond formation is controlling [Hammer and Lauffenburger, 1987]. Most physiologic situations of cell capture from the flowing bloodstream by blood vessel endothelium appear to be under kinetic control. That is, the crucial factor is the rate at which the first receptor-ligand bond can be formed; apparently the bond strength is sufficiently great to hold the cell at the vessel wall despite the distracting force due to fluid flow. Rolling of neutrophils along the endothelium is mediated by members of the selectin family, slowing the cells down for more stable adhesion and extravasation mediated by members of the integrin family. It is not yet clear what special property of the selectins accounts for their ability to yield rolling behavior; one candidate is a combination of both fast association and dissociation rate constants [Lawrence and Springer, 1991], whereas another is a low compliance permitting the dissociation rate constant to be relatively unaffected by stress [Hammer and Apte, 1992]. It will be important to determine what the key properties are for proper design of capture-enhancing or -inhibiting materials or regimens.

14.4 Cell Migration

Movement of cells across or through two- and three-dimensional substrata, respectively, is governed by receptor-ligand interactions in two distinct ways. First, active migration arising from intracellular motile forces requires that these forces can be effectively transmitted to the substratum as differential traction. Adhesion receptors, mainly of the integrin superfamily, are involved in this process. Integrin interactions with their extracellular matrix ligands can influence the linear cell translocation speed. Second, the intracellular motile forces must be stimulated under the specific conditions for which movement is desired. Often, the ligands that stimulate cytoskeletally generated motile forces can do so in a spatially dependent fashion; i.e., cells can be induced to extend membrane lamellipodia preferentially in the direction of higher ligand concentrations, leading to net movement up a ligand concentration gradient. These ligands are termed *chemotactic attractants*; many such attractants can additionally modulate movement speed and are thus also termed *chemokinetic agents*. Chemotactic and chemokinetic ligands include some growth factors as well as other factors seemingly more dedicated to effects on migration.

Two different types of intracellular motile forces are involved, one that generates membrane lamellipod extension and one that generates cell body contraction; it is the latter force that must be transmitted to the substratum as traction in order for locomotion to proceed. Indeed, it must be transmitted in a spatially asymmetrical manner so that attachments formed at the cell front can remain adhered, while attachments at the cell rear are disrupted. Since the contraction force is likely to be isotropic in nature, this asymmetry must arise at the site of the force-transmission linkage. Migration speed of a particular cell type on a given ligand-coated surface thus depends on (1) generation of intracellular motile forces leading to lamellipod extension and to subsequent cell body contraction, stimulated by chemokinetic and/or chemotactic ligand binding to corresponding receptors or by adhesion receptor binding to appropriate soluble or substratum-bound ligands, (2) an appropriate balance between intracellular contractile force and overall level of cell-substratum traction, and (3) a significant front versus rear difference in cell-substratum traction. Speed should be roughly proportional to the first and last quantities but dependent on the middle quantity in biphasic fashion — maximal at an intermediate level of motile force-to-traction-force ratio [DiMilla et al., 1991].

Since many adhesion ligands stimulate intracellular motile force generation, an important parameter for design purposes is thus the effective density at which they are present on the substratum. Migration should be maximized at an intermediate density at which the cell contraction force is roughly in balance with the cell-substratum adhesiveness, consistent with observations for both two-dimensional [DiMilla et al., 1993] and three-dimensional [Parkhurst and Saltzman, 1992] movement environments. Since adhesive force is related to the ligand-receptor binding affinity, use of ligands with different affinities provides a means for migration under conditions of different ligand densities. Recalling also that the degree of receptor organization into aggregates can dramatically increase cell-substratum adhesiveness, use of ligands yielding different degrees of receptor aggregation can similarly influence migration speed. Engineering attempts to develop useful substrata for cell migration, e.g., for wound-healing scaffolding or tissue-regeneration biomaterials also need to consider possible effects of microarchitecture, i.e., the physical geometry of three-dimensional matrices through which the cells may crawl and ligands or nutrients may diffuse [Cima and Langer, 1993].

Addition of soluble receptor-binding ligand competitors into the extracellular medium is another approach to affecting cell migration speed by means of altered cell-substratum adhesiveness. As the concentration of a competitor is increased, adhesiveness should decrease. However, since migration speed varies with adhesiveness in biphasic manner, the result of adding the competitor can be either a decrease or an increase in motility, depending on whether the adhesiveness in the absence of the competitor is below or above optimal [Wu et al., 1994]. For example, the rate of blood capillary formation, which is related to the speed of microvessel endothelial cell migration, can be either diminished or enhanced by addition of soluble integrin-binding competitors [Nicosia and Bonnano, 1991; Gamble et al., 1993].

Finally, migration speed may be altered genetically, e.g., by changing the expression level of an adhesion receptor [Schreiner et al., 1989]. An increase in the number of receptors may lead to either an increase or a decrease in motility, depending on whether the change in adhesiveness is helpful or detrimental. At the same time, signaling for motile force generation, if it occurs through the same receptor, will provide an accompanying positive influence. Receptors may be altered in their ligand-binding or cytoskeleton-coupling capabilities as well, providing alternative approaches for affecting force transmission.

Attempts to manipulate the direction of cell migration fall into three categories: soluble chemotactic ligands, immobilized haptotactic ligands, and contact guidance. *Haptotaxis* is commonly referred to as directed migration on a density gradient of an immobilized ligand, regardless of whether it is due to differential adhesiveness or to spatially dependent membrane extension. If the latter effect is causative, the phenomenon should be more precisely termed *chemotaxis* but merely with an immobilized ligand; the classic definition of haptotaxis requires an adhesion effect. Examples in which efforts have been made to manipulate the direction of cell migration including homing of cytotoxic lymphocytes to tumor sites, neovascularization of healing wound tissue, and neuron outgrowth.

Contact guidance can be produced using either patterned substrata, in which pathways for cell migration are delineated by boundaries of adhesiveness [Kleinfeld et al., 1988], or by a alignment of matrix fibers for directional orientation of cell movement [Dickinson et al., 1994]. A useful technology for generation of ligand concentration gradients is controlled release of chemotactic ligands from polymer matrices [Edelman et al., 1991]. Gradients required for effective cell guidance have been found to be approximately a few percent in concentration across a cell length, and they must persist for many multiples of the cell mean path length; this is the product of the persistence time (i.e., the mean time between significant direction changes) and the linear speed. Analysis has predicted that the affinity of a chemotactic ligand or, more precisely, the dissociation rate constant may be a key parameter in governing the sensitivity of a cell to a chemotactic ligand concentration gradient [Tranquillo et al., 1988]. Data on directional responses of neutrophil leukocytes to gradients of interleukin 8 show that these cells respond more sensitively to variants of IL-8 possessing lower binding affinities [Clark-Lewis et al., 1991]. Although this finding is consistent with theory, involvement of multiple forms of the IL-8 receptor with possibly varying signaling capabilities also must be considered as an alternative cause. Work with platelet-derived growth factor as a chemotactic attractant has elucidated features of the PDGF receptor involved in sensing ligand gradients, permitting approaches for pharmacologic or genetic manipulation of chemotactic receptors to modify directional migration responses [Kundra et al., 1994].

References

Buckley A, Davidson JM, Kamerath CD, et al. 1985. Sustained release of epidermal growth factor accelerates wound repair. *Proc Natl Acad Sci USA* 82:7340.

Buschmann MD, Gluzband YA, Grodzinsky AJ, et al. 1992. Chondrocytes in agarose culture synthesize a mechanically functional extracellular matrix. *J Orthop Res* 10:745.

Cima LG. 1994. Polymer substratas for controlled biological interactions. *J Coll Biochem* 56:155.

Cima LG, Vacanti JP, Vacanti C, et al. 1991. Tissue engineering by cell transplantation using degradable polymer substrates. *ASME Trans J Biomec Eng* 113:143.

Cima LG, Langer R. 1993. Engineering human tissue. *Chem Eng Progr* June:46.

Clark-Lewis I, Schumacher C, Baggiolini M, Moser B. 1991. Structure-activity relationships of IL-8 determined using chemically synthesized analogs. *J Biol Chem* 266:23128.

Darnell J, Lodish H, Baltimore D. 1986. *Molecular Cell Biology*. New York, Scientific American Books.

Dickinson RB, Guido S, Tranquillo RT. 1994. Correlation of biased cell migration and cell orientation for fibroblasts exhibiting contact guidance in oriented collagen gels. *Ann Biomed Eng* 22:342.

DiMilla PA, Barbee K, Lauffenburger DA. 1991. Mathematical model for the effects of adhesion and mechanics on cell migration speed. *Biophys J* 60:15.

DiMilla PA, Stone JA, Quinn JA, et al. 1993. An optimal adhesiveness exists for human smooth muscle cell migration on type-IV collagen and fibronectin. *J Cell Biol* 122:729.

Edelman ER, Mathiowitz E, Langer R, Klagsbrun M. 1991. Controlled and modulated release of basic fibroblast growth factor. *Biomaterials* 12:619.

French AR, Sudlow GP, Wiley HS, Lauffenburger DA. 1993. Postendocytic trafficking of EGF-receptor complexes is mediated through saturable and specific endosomal interactions. *J Biol Chem* 269:15749.

Gamble JR, Mathias LJ, Meyer G, et al. 1993. Regulation of in vitro capillary tube formation by anti-integrin antibodies. *J Cell Biol* 121:931.

Hammer DA, Apte SA. 1992. Simulation of cell rolling and adhesion on surfaces in shear flow: General results and analysis of selectin-mediated neutrophil adhesion. *Biophys J* 63:35.

Hammer DA, Lauffenburger DA. 1987. A dynamical model for receptor-mediated cell adhesion to surfaces. *Biophys J* 52:475.

Hubbell JA, Massia SP, Desai NP, Drumheller PD. 1991. Endothelial cell-selective materials for tissue engineering in the vascular graft via a new receptor. *Biotechnology* 9:568.

Ingber DE. 1993. The riddle of morphogenesis: a question of solution chemistry or molecular cell engineering? *Cell* 75:1249.

Ingber DE, Madri JA, Folkman J. 1987. Endothelial growth factors and extracellular matrix regulate DNA synthesis through modulation of cell and nuclear expansion. *In Vitro Cell Dev Biol* 23:387.

Kelly JJ, Koury MJ, Bondurant MC, et al. 1993. Survival or death of individual proerythroblasts results from differing erythropoietin sensitivities: A mechanism for controlled rate of erythrocyte production. *Blood* 82:2340.

Kleinfeld D, Kahler KH, Hockberger PE. 1988. Controlled outgrowth of dissociated neurons on patterned substrates. *J Neurosci* 8:4098.

Knauer DJ, Wiley HS, Cunningham DD. 1984. Relationship between epidermal growth factor receptor occupancy and mitogenic response: Quantitative analysis using a steady-state model system. *J Biol Chem* 259:5623.

Kundra V, Escobedo JA, Kazlauskas A, et al. 1994. Regulation of chemotaxis by the PDGF-receptor β. *Nature* 367:474.

Kuo SC, Lauffenburger DA. 1993. Relationship between receptor/ligand binding affinity and adhesion strength. *Biophys J* 65:2191.

Lauffenburger DA, Aebischer P. 1993. Cell and tissue engineering: Overview. In *Research Opportunities in Biomolecular Engineering: The Interface Between Chemical Engineering and Biology*. US Dept of Health and Human Services Administrative Document, pp 109–113.

Lauffenburger DA, Lindermann JJ. 1993. *Receptors: Models for Binding, Trafficking, and Signaling.* New York, Oxford University Press.

Lotz MM, Burdsal CA, Erickson HP, McClay DR. 1989. Cell adhesion to fibronectin and tenascin: Quantitative measurements of initial binding and subsequent strengthening response. *J Cell Biol* 109:1795.

Mahama P, Lindermann JJ. 1994. Monte Carlo study on the dynamics of G-protein activation. *Biophys J* 67:1345.

Massia SP, Hubbell JA. 1991. An RGD spacing of 440 nm is sufficient for integrin $\alpha_v\beta_3$-mediated fibroblast spreading and 140 nm for focal contact and stress fiber formation. *J Cell Biol* 114:1089.

Nerem RM. 1991. Cellular engineering. *Ann Biomed Eng* 19:529.

Nicosia RF, Bonanno E. 1991. Inhibition of angiogenesis in vitro by RGD-containing synthetic peptide. *Am J Pathol* 138:829.

Parkhurst M. Saltzman WM. 1992. Quantification of human neutrophil motility in three-dimensional collagen gels: Effect of collagen concentrations. *Biophys J* 61:306.

Powell EM, Sobarzo MR, Saltzman WM. 1990. Controlled release of nerve growth factor from a polymeric implant. *Brain Res* 515:309.

Reddy CC, Wells A, Lauffenburger DA. 1993. Proliferative response of fibroblasts expressing internalization-deficient EGF receptors is altered via differential EGF depletion effects. *Biotech Progr* 10:377.

Renner WA, Hatzimanikatis V, Eppenberger HM, Bailey JE. 1993. Recombinant cyclin E overexpression enables proliferation of CHO K1 cells in serum- and protein-free medium. Preprint from Institute for Biotechnology, ETH-Hoenggerberg, Zurich, Switzerland.

Robb RJ. 1982. Human T-cell growth factor: Purification, biochemical characterization, and interaction with a cellular receptor. *Immunobiology* 161:21.

Schreiner CL, Bauer JS, Danilov YN, et al. 1989. Isolation and characterization of chinese hamster ovary cell variants deficient in the expression of fibronectin receptor. *J Cell Biol* 109:3157.

Schultz GS, White M, Mitchell R, et al. 1987. Epithelial wound healing enhanced by TGFα and VGF. *Science* 235:350.

Singhvi R, Kumar A, Lopez GP, et al. 1994. Engineering cell shape and function. *Science* 264:696.

Tranquillo RT, Lauffenburger DA, Zigmond SH. 1988. A stochastic model for leukocyte random motility and chemotaxis based on receptor binding fluctuations. *J Cell Biol* 106:303.

Tranquillo RT, Durrani MA, Moon AG. 1992. Tissue engineering science: Consequences of cell traction force. *Cytotechnology* 10:225.

Ward MD, Hammer DA. 1993. A theoretical analysis for the effect of focal contact formation on cell-substrate attachment strength. *Biophys J* 64:936.

15

Metabolic Engineering

Craig Zupke
Massachusetts General Hospital
and the Shriners Burns Institute

Metabolic engineering can be defined as the modification of cellular metabolism to achieve a specific goal. Metabolic engineering can be applied to both prokaryotic and eukaryotic cells, although applications to prokaryotic organisms are much more common. The goals of metabolic engineering include improved production of chemicals endogenous to the host cell, alterations of the substrate required for growth and product synthesis, synthesis of new products foreign to the host cell, and addition of new detoxification activities. As metabolic engineering has evolved, some general principles and techniques have emerged. In particular, developments in genetic engineering enable highly specific modification of cellular biochemical pathways. However, the complexity of biologic systems and their control severely limits the ability to implement a rational program of metabolic engineering. Early attempts at metabolic engineering were based on a simplistic analysis of cellular biochemistry and did not live up to expectations [Stephanopoulos and Sinskey, 1993]. An important realization is that metabolism must be considered as a network of interrelated biochemical reactions. The concept of a single rate-limiting step usually does not apply — instead, the control of a metabolic pathway is frequently distributed over several steps. The choice of specific modifications to achieve a desired goal is difficult because of this distributed nature of control. These issues are especially relevant in medical applications of metabolic engineering, where there is the added complication of interacting organ systems.

The field of metabolic engineering is still young, and advancements are continuing, especially in the analysis of metabolic networks and their control. Recent advances include techniques for the experimental determination of kinetic or control parameters necessary to apply established mathematical formalisms to metabolic networks. Coupled with computer optimization and simulation, these techniques should lead to a greater success rate for the directed modification of cellular metabolism. In addition, the knowledge and experience gained from engineering cellular metabolism can be applied to individual organs or to the whole body to help guide the development and evaluation of new drugs and therapies.

15.1 Basic Principles

Metabolic engineering is inherently multidisciplinary, combining knowledge and techniques from cellular and molecular biology, biochemistry, chemical engineering, mathematics, and computer science. The basic cycle of metabolic engineering is summarized in the flow sheet in Fig. 15.1. The first step in the process is to define the problem or goal to which metabolic engineering will be applied. The next step is to analyze the relevant biochemistry to determine the specific modifications that will be attempted. Recombinant DNA techniques are then used to implement the modifications, and the analysis of their impact completes one pass through the metabolic engineering cycle. This process is continued in an iterative fashion until the desired result is achieved. In the discussion that follows, each step of the cycle will be addressed individually.

15.2 Problem Definition

As mentioned above, the typical goals of metabolic engineering include improving the production of endogenous chemicals, altering the organism substrate requirements, synthesizing new products, and adding new detoxification activities. There are a variety of ways that these goals can be achieved through genetic engineering and mutation and selection. Through genetic engineering, heterologous enzymes can be added to a cell, either singly or in groups. The addition of new enzymatic activities can be used to meet any of the metabolic engineering goals. Mutation and selection can be used to affect the regulation of particular enzymes and thus are most useful for altering metabolite flow.

Addition of New Activities

The distinction between product formation, substrate utilization, and detoxification is simply a matter of emphasizing different aspects of a metabolic pathway. In product formation, the end of a metabolic pathway is the focal point. In substrate utilization and detoxification, it is the input to the pathway that is important. With detoxification, there is the additional requirement that the toxin is converted to less harmful compounds, but what they are is not important. In the discussion that follows, the addition of new activity for the synthesis of new products will be emphasized. However, since they all involve adding new activity to a metabolic pathway, the same techniques used for the synthesis of new products also can be used to alter substrate requirements or add new detoxification activity.

There are several possible strategies for the synthesis of new products. The choice of strategy depends on the biochemistry present in the host organism and the specific product to be synthesized. In many instances, pathway completion can be used, which only involves the addition of a few enzymes that catalyze the steps missing from the existing metabolic pathway. Entire biochemical pathways also can be transferred to a heterologous host. The motivation for this type of transfer is that the host organism may be more robust, have a more desirable substrate requirement, or have inherently higher productivities than the organism that naturally possesses the biochemical pathway of interest. The transfer of multistep pathways to heterologous hosts has been used frequently for the production of antibiotics because of the clustering of the genes involved [Malpartida, 1990]. When the genes are clustered, they can be cloned and inserted into the host organism as a single unit.

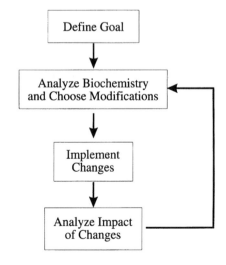

FIGURE 15.1   The metabolic engineering cycle.

The procedures described above involve the transfer of a known metabolic activity from one organism to another. It is also possible to create completely new products, because many enzymes can function with multiple substrates. If an enzyme is introduced into a cell and is thus presented with a metabolite similar to its normal substrate but which was not present in the donor cell, a completely new reaction may be catalyzed. Similarly, if a foreign enzymatic activity produces an intermediate foreign to the host cell, that intermediate may undergo further reactions by endogenous enzymes and produce a novel product. This phenomenon has been used to synthesize novel antibiotics in recombinant *Streptomyces* [McAlpine, 1987; Epp, 1989] and to engineer an *Escherichia coli* for the degradation of trichlorethylene [Winter, 1989]. Of course, the production of new products also may be an undesirable side effect of the introduction of a heterologous enzyme and could interfere with the production of the desired product. The complexity of enzyme activities makes the prediction of negative side effects extremely difficult.

Improving Existing Metabolism

Although a cell or organism may possess a particular metabolic pathway, the flow of material through that pathway (metabolic flux) may be suboptimal. Both recombinant DNA techniques and classic mutation-selection techniques can help to redirect metabolic fluxes. The principal ways to redirect metabolic fluxes are by increasing or decreasing the activity of specific enzymes and by changing their regulation. The choice of which enzymes and how to alter them can be very complicated because of the distributed nature of metabolic control. However, some basic concepts can be illustrated by the analysis of an idealized metabolic network for threonine synthesis, as shown in Fig. 15.2. Aspartate is the biosynthetic precursor for threonine, as well as for lysine, methionine, and isoleucine. If our goal is to maximize the production of threonine, then we may want to decrease the activity of the enzymes that catalyze the production of the other amino acids derived from aspartate. This would result in less of the carbon flow from aspartate being drawn off by the unwanted amino acids, and thus a larger flux to threonine might be achieved. An alternative way of increasing the flux to threonine is to amplify the activity of the enzymes along the pathway from aspartate to threonine. Finally, accumulation of the desired product, threonine, will lead to the reduction of flux from aspartate to ASA and from ASA to homoserine because of feedback inhibition. This would result in a reduction in threonine synthesis as well. If the feedback inhibition were removed, threonine could then accumulate without affecting its rate of synthesis. In practice, multiple modifications would probably be required to achieve the desired goal.

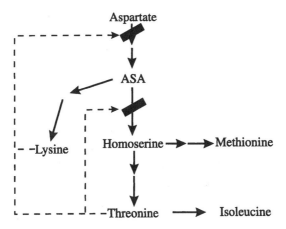

FIGURE 15.2   A simple metabolic network for threonine synthesis. The solid arrows represent enzyme-catalyzed biochemical reactions. The dotted arrows indicate allosteric inhibition (ASA = aspartate semialdehyde).

15.3 Analysis of Metabolic Networks

Metabolic Flux Analysis

The measurement or estimation of metabolic fluxes is useful for the evaluation of metabolic networks and their control. Direct measure of intracellular fluxes is possible with *in vivo* nuclear magnetic resonance (NMR) spectroscopy. However, the inherent insensitivity of NMR limits its applicability. A more general strategy combines material balances with carefully chosen measurements to indirectly estimate metabolic fluxes. In isotopic tracer methods, a material balance is performed on the labeled isotope, and the measured isotope distribution is used to estimate the fluxes of interest. Alternatively, a total mass or carbon balance can be combined with extensive measurements of the rates of change of extracellular metabolic concentrations to calculate intracellular fluxes. Used individually or together, these methods can provide a great deal of information about the state of a biochemical system.

In isotope tracer methods, the cells to be studied are provided with a substrate specifically labeled with a detectable isotope (usually ^{14}C or ^{13}C). The incorporation of label into cellular material and by-products is governed by the fluxes through the biochemical pathways. The quantity and distribution of label are measured and combined with knowledge of the biochemistry to determine the intracellular fluxes. The choices of substrate labeling patterns, as well as which by-products to measure, are guided by careful analysis of the assumed biochemical network. These experiments are usually performed at isotopic steady state so that the flow of isotope into each atom of a metabolite equals the flux out. For the nth atom of the kth metabolite, the flux balance is [Blum, 1982]:

$$\sum_i V_{i,k} S_i(m) = S_k(n) \sum_o V_{o,k} , \tag{15.1}$$

where $S_i(m)$ is the specific activity of the mth atom of the ith metabolite that contributes to the labeling of the nth atom of the kth metabolite, $S_k(n)$, and $V_{i,k}$ and $V_{o,k}$ are the input and output fluxes of the kth metabolite, respectively. The system of equations represented by Eq. (15.1) can be solved for the fluxes $(V_{i,k}, V_{o,k})$ from measurements of the specific activities. A similar analysis can be applied to the isotope isomers, or isotopomers, yielding more information from a single experiment. Isotopomer analysis is not just concerned with the average enrichment of individual atoms. Instead, it involves quantifying the amounts of the different isotopomers that occur at a specific metabolic state. The isotopomer distribution contains more information than that obtained from positional enrichments, making it generally a more powerful technique. Both mass spectrometry and NMR can be used to analyze isotopomer distributions.

Flux estimation based only on material balances is identical, in principle, to flux estimation using isotopic tracers and has been applied to many biochemical processes, including lysine synthesis [Vallino, 1989] and rat heart metabolism [Safer and Williamson, 1973]. Instead of following the fate of specific atoms (i.e., ^{13}C or ^{14}C), all are considered equally through the measurement of the rates of change of substrates and products. The analysis is simplified in some ways because only the stoichiometry of the biochemical reactions is important, and no knowledge of the chemical mechanisms is needed. Some resolving power is lost, however, with the loss of the ability to determine the source of individual reactions. The analysis is usually formulated as a matrix equation:

$$\mathbf{r} = \mathbf{A}\mathbf{x} , \tag{15.2}$$

where $\mathbf{r}$ is a vector of metabolite rates of change, $\mathbf{A}$ is the matrix of stoichiometric coefficients for the biochemical network, and $\mathbf{x}$ is a vector of the metabolic fluxes. If the rates of change of the metabolites are measured, then Eq. (15.2) can be solved for the flux vector $\mathbf{x}$ by the method of least squares:

$$\mathbf{x} = \left(\mathbf{A}^T \mathbf{A}\right)^{-1} \mathbf{A}^T \mathbf{r} . \tag{15.3}$$

Typically, extracellular metabolites are measured, while intracellular metabolites are assumed to be at pseudo-steady state and their rates of change are taken to be zero.

Metabolic Control Analysis

A *biochemical network* is a system of enzyme-catalyzed reactions that are interconnected through shared metabolites. At steady state, the fluxes through each pathway will be a function of the individual enzyme kinetic properties, as well as the network architecture. The activity of a particular enzyme will affect the concentration of its reactants and products and thus will influence the flux through pathways upstream and downstream. Metabolic control analysis (MCA) grew from the work originally presented by Kacser and Burns [1973] and Heinrich and Rapoport [1977]. MCA provides a framework for analyzing and quantifying the distributed control that enzymes exert in biochemical networks.

In the discussion that follows, a metabolic network consists of enzymes e, metabolites X, substrates S, and products P. For simplicity, we assume that the reaction rate v_i is proportional to the enzyme concentration e_i. If this is not true, then some modifications of the analysis are required [Liao and Delgado, 1993]. The concentrations of the products and substrates are fixed, but the metabolite concentrations are free to change in order to achieve a steady-state flux J. An example of a simple, unbranched network is

$$S \to X_1 \to X_2 \to X_3 \to P$$
$$e_1 \quad e_2 \quad e_3 \quad e_4$$

MCA is essentially a sensitivity analysis that determines how perturbations in a particular parameter (usually enzyme concentration) affect a variable (such as steady-state flux). The measures of the sensitivities are control coefficients defined as follows:

$$C_{e_i}^J \equiv \frac{e_i}{J} \frac{\partial J}{\partial e_i} = \frac{\partial \ln |J|}{\partial \ln e_i}. \tag{15.4}$$

The flux control coefficient C_{ei}^j is a measure of how the flux J changes in response to small perturbations in the concentration or activity of enzyme i. The magnitude of the control coefficient is a measure of how important a particular enzyme is in the determination of the steady-state flux. The summation theorem relates the individual flux control coefficients:

$$\sum_i C_{e_i}^J = 1. \tag{15.5}$$

A large value for C_{ei}^j indicates that an increase in the activity of enzyme $i(e_i)$ should result in a large change in the metabolic flux. Thus enzymes with large flux control coefficients may be good targets for metabolic engineering. Usually, there are several enzymes in a network with comparably large C values, indicating that there is no single rate-limiting enzyme.

The challenge in analyzing a metabolic network is determination of the flux control coefficients. It is possible to determine them directly by "enzyme titration" combined with the measurement of the new steady-state flux. For *in vivo* systems, this would require alteration of enzyme expression through an inducible promoter and is not very practical for a moderately sized network. A more common method of altering enzyme activities involves titration with a specific inhibitor. However, this technique can be complicated by nonspecific effects of the added inhibitor and unknown inhibitor kinetics [Liao and Delgado, 1993].

Although the control coefficients are properties of the network, they can be related to individual enzyme kinetics through elasticity coefficients. If a metabolic concentration X is altered, there will be an effect on the reaction rates in which X is involved. The elasticity coefficient ε_X^{vi} is a measure of the effect changes in X have on v_i, the rate of reaction catalyzed by enzyme e_i:

$$\epsilon_X^{v_i} \equiv \frac{X}{v_i}\frac{\partial v_i}{\partial X} = \frac{\partial \ln v_i}{\partial \ln X} \tag{15.6}$$

The flux control connectivity theorem relates the flux control coefficients to the elasticity coefficients:

$$\sum_i C_{e_i}^J \epsilon_{X_k}^{v_i} = 0 \qquad \text{for all } k \tag{15.7}$$

In principle, if the enzyme kinetics and steady-state metabolite concentrations are known, then it is possible to calculate the elasticities and through Eq. (15.7) determine the flux control coefficients.

Flux Control Coefficients from Transient Metabolite Concentrations

Recent efforts to address the difficulty of determining MCA parameters include the use of transient measurements of metabolite concentrations to give good estimates of the flux control coefficients [Delgado and Liao, 1992]. The key assumption in the dynamic approach is that the reaction rates are reasonably linear around the steady state. When this is true and a transient condition is induced (by changing substrate or enzyme concentration), the transient fluxes $v_i(t)$ are constrained by the following equation:

$$\sum_{i=1}^{r} C_{e_i}^j v_i(t) = J, \tag{15.8}$$

where r equals the number of reactions in the network. Measurements of transient metabolite concentration allow the determination of regression coefficients α_i from the following equation:

$$\sum_{i=1}^{n} \alpha_i \left[X_i(t) - X_i(0) \right] = 0, \tag{15.9}$$

where n equals the number of metabolites measured. The flux control coefficients are then determined from

$$\left[C_{e_1}^J \quad C_{e_2}^J \quad \cdots \quad C_{e_r}^J \right] = J \left[\alpha_1 \quad \alpha_2 \quad \cdots \quad \alpha_n \right] \mathbf{A}, \tag{15.10}$$

where $\mathbf{A}$ is the matrix of stoichiometric coefficients. Using this method requires the measurements of transient metabolite concentrations, which are used to calculate the α_i values, and the accumulation or depletion of external metabolites to determine the steady-state flux J. The measurement of transient metabolite concentrations can be difficult but is possible with *in vivo* NMR or from cell extracts taken at different time points.

"Top-Down" MCA

The traditional approach of MCA can be considered to be "bottom-up," since all the individual enzyme flux control coefficients are determined in order to describe the control structure of a large network. The "top-down" approach makes extensive use of lumping of reactions together to determine group flux control coefficients [Brown, 1990]. These can give some information about the overall control of a metabolic network without its complete characterization.

Consider a simple, multireaction pathway:

$$S \rightarrow \rightarrow \rightarrow X \rightarrow \rightarrow \rightarrow P$$

$$\text{produces } X \quad \text{consumes } X \cdot$$

$$J_1 \qquad\qquad J_2$$

The reactions of a metabolic network are divided into two groups, those which produce a particular metabolite X and those which consume it. By manipulating the concentration X and measuring the resulting fluxes J_1 and J_2 "group" or "overall" elasticities $*\varepsilon$ of the X producers and X consumers can be determined. Application of the connectivity theorem (Eq. 15.7) then permits the calculation of the group control coefficient for both groups of reactions. Each pathway can subsequently be divided into smaller groups centered around different metabolites, and the process repeated. The advantage of the top-down approach is that useful information about the control architecture of a metabolic network can be obtained more quickly. This approach is particularly appropriate for highly complex systems such as organ or whole-body metabolism.

Large Deviations

One of the limitations of MCA is that it only applies when the perturbations from the steady state are small. Experimentally, it is much easier to induce large changes in enzyme or metabolite concentrations, and in terms of metabolic engineering, the desired perturbations are also likely to be large. Small and Kacser [1993a, 1993b] have developed an analysis based on large deviations and related it to MCA.

In this discussion, e_i^0 is the original concentration of enzyme i, $e_i^r = re_i^0$, where r is noninfinitesimal. Thus e_i^r represents a large perturbation to the system. J^0 is the flux at the original steady state, and J^r is the flux after the large perturbation. A deviation index D is used to characterize the change from the original steady state:

$$D_{e_i}^{J^r} = \left(\frac{\Delta J}{\Delta e_i} \right) \frac{e_i^r}{J^r} , \tag{15.11}$$

where $\Delta J = J^r - J^0$ and $\Delta e_i = e_i^r - e_i^0$. By assuming that each individual enzymatic reaction rate is a linear function of the participating metabolites, it can be shown that the deviation index and control coefficients are equivalent:

$$D_{e_i^r}^{J^r} = C_{e_i}^{J_0^0} . \tag{15.12}$$

Similarly, an alternate deviation index $*D$ can be defined as

$$*D_{e_i^r}^{J^r} = \left(\frac{\Delta J}{\Delta e_i} \right) \frac{e_i^0}{J^0} , \tag{15.13}$$

and it is equivalent to the control coefficient at the new steady state:

$$D_{e_i^r}^{J^r} = C_{e_i^r}^{J^r} . \tag{15.14}$$

Thus, with a single large perturbation, the control coefficients at the original and new steady states can be estimated. This analysis has been extended to branched pathways, but relationships between the deviation indices and the flux control coefficients are more complicated and depend on the magnitude of the deviation (r). If the subscript a designates one branch and b another, the following relationship holds:

$$D_{e_a^r}^{J_b^r} = C_{e_a^0}^{J_b^0} \frac{1}{1 - \left(C_{e_a^0}^{J_a^0} - C_{e_a^0}^{J_b^0} \right) \dfrac{r-1}{r}} . \tag{15.15}$$

Metabolic control analysis, especially with the recent innovations for determining flux control coefficients, is a powerful tool for the analysis of metabolic networks. It can describe the control architecture of a biochemical reaction network and identify which steps are the most promising targets for efforts at metabolic engineering.

Pathway Synthesis

The diversity of biochemical reactions found in nature is quite extensive, with many enzymes being unique to a particular organism. Through metabolic engineering, it is possible to construct a metabolic network that performs a specific substrate-to-product transformation not found in nature. When exploring the possibility of synthesizing new biochemical pathways, there are several key issues that must be addressed. Given a database of possible enzymatic activities and a choice of substrate and product, one must first generate a complete set of possible biochemical reactions that can perform the desired conversion. Once they are generated, the set of possible biochemical pathways must be checked for thermodynamic feasibility and evaluated in terms of yields, cofactor requirements, and other constraints that might be present. In addition, the impact of an engineered metabolic pathway on the growth and maintenance of the host cell is also important to evaluate.

The problem of synthesizing a complete set of possible biochemical pathways subject to constraints on allowable substrates, intermediates, and by-products has been solved [Mavrovouniotis, 1990]. A computer algorithm allows the efficient determination of a complete and correct set of biochemical pathways that connect a substrate to a product. In addition, a complementary computer algorithm evaluates metabolic pathways for thermodynamic feasibility [Mavrovouniotis, 1993]. The key concept in the thermodynamic analysis is that evaluation of the feasibility of biologic reactions requires the specification of the concentrations of the products and reactants. The standard free-energy change ΔG^0 is not sufficient because physiologic conditions are significantly different from standard conditions. Both local and distributed thermodynamic bottlenecks can be determined by incorporating knowledge of the metabolite concentration ranges. This thermodynamic analysis of a biochemical pathway can pinpoint specific reactions or groups of reactions that should be modified or bypassed in order to better favor product formation.

The addition of new biochemical pathways or the modification of existing pathways is likely to affect the rest of the cellular metabolism. The new or altered pathways may compete with other reactions for intermediates or cofactors. To precisely predict the impact of a manipulation of a metabolic network is virtually impossible, since it would require a perfect model of all enzyme kinetics and of the control of gene expression. However, with relatively simple linear optimization techniques, it is possible to predict some of the behavior of a metabolic network. The procedure involves the solution of Eq. (15.2) for the metabolic fluxes for networks that have more unknowns **x** than knowns **r**. An undetermined system of linear equations can be solved uniquely using linear optimization techniques if a "cellular objective function" is postulated. Examples of objective functions used in the literature include minimizing ATP or NADH production [Savinell, 1992] and maximizing growth or product formation [Varma, 1993]. Examination of the fluxes gotten from the linear optimization can indicate potential effects of a proposed metabolic change. For example, the maximum growth of *E. coli* was found to decrease in a piecewise linear fashion as leucine production increased [Varma, 1993]. In principle, linear optimization also could be applied to whole-body metabolism to evaluate the effects of metabolically active drug or genetic therapies.

15.4 Implementing Changes

The techniques available to implement specific changes in both eukaryotic and prokaryotic cells are quite powerful. Through classic mutation-selection and modern genetic engineering, it is possible to amplify or attenuate existing enzyme activity, add completely new activities, and modify the regulation of existing pathways. A detailed description of all the genetic engineering techniques available is beyond the scope of this discussion, and only a general overview will be presented.

Mutation-Selection

Mutation and selection constitute a method of manipulating the phenotype of a cell. Cells are exposed to a mutagenic environment and then placed in a selective medium in which only those cells that have a desired mutation can grow. Alternatively, cells can be screened by placing them in a medium in which it is possible to visibly detect those colonies that possess the desired phenotype. Although it is a random process, careful design of selection or screening media can result in alterations in specific enzymes. For example, growth in a medium that contains an allosteric inhibitor may be the result of the loss of the allosteric inhibition. By altering metabolite concentrations or adding substrate analogues, it is possible to select for increased or decreased activity or for changes in regulation. The major drawback of mutation and selection for metabolic engineering is the lack of specificity. There may be multiple ways that a particular phenotype can be generated, so there is no guarantee that the enzyme that was targeted was affected at all.

Recombinant DNA

Recombinant DNA techniques are very flexible and powerful tools for implementing specific metabolic changes. The cloning of genes encoding a specific enzyme is relatively routine, and the subsequent insertion and expression are also straightforward for many host cells. With the choice of a very active promoter, the activity of a cloned enzyme can be greatly amplified. In addition, insertion of a heterologous activity into a host cell can serve to alter the regulation of a metabolic pathway, assuming the foreign enzyme has different kinetics than the original. Enzyme activity also can be attenuated by antisense sequences that produce mRNA complementary to the endogenous mRNA and thus form double-stranded RNA complexes that cannot be translated. Finally, heterologous recombination can be used to completely remove a particular gene from the host genome.

15.5 Analysis of Changes

The techniques used to analyze the biochemistry at the beginning of the metabolic engineering cycle are also applicable for assessing the effects of any attempted changes. The effect of the specific change can be evaluated, as well as its impact on the activity of the whole metabolic network. It is quite possible for a desired change to be made successfully at the enzyme level but not have the desired effect on the metabolic network. If a change was partially successful, then the resulting cells can be put through another iteration of metabolic engineering in order to make further improvements. If, on the other hand, a particular change was not successful, then any information which that failure gives about the regulation and control of the metabolic network can be used to make another attempt at implementing the desired change.

15.6 Summary

Metabolic engineering is an evolving discipline that tries to take advantage of the advances that have been made in the genetic manipulation of cells. Through genetic engineering we have the capability of making profound changes in the metabolic activity of both prokaryotic or eukaryotic cells. However, the complexity of the regulation of biochemical networks makes the choice of modifications difficult. Recent advances in the analysis of metabolic networks via MCA or other techniques have provided some of the tools needed to implement a rational metabolic engineering program. Finally, metabolic engineering is best viewed as an iterative process, where attempted modifications are evaluated and the successful cell lines improved further.

Defining Terms

Flux: The flow of mass through a biochemical pathway. Frequently expressed in terms of moles of metabolite or carbon per unit time.

Heterologous enzyme: An enzyme from a foreign cell that has been expressed in a host cell. Frequently used to provide activity normally not present or to alter the control structure of a metabolic network.

Host: The cell that has had foreign DNA inserted into it.

Isotopomers: Isomers of a metabolite that contain different patterns of isotopes (for metabolic studies, the carbon isotopes ^{12}C, ^{13}C, and ^{14}C are the most commonly analyzed).

Metabolic engineering: The modification of cellular metabolism to achieve a specific goal. Usually performed with recombinant DNA techniques.

Metabolic network: A system of biochemical reactions that interact through shared substrates and allosteric effectors.

Pathway completion: The addition of a small number of heterologous enzymes to complete a biochemical pathway.

References

Blum JJ, Stein RB. 1982. On the analysis of metabolic networks. In RF Goldberger, KR Yamamoto (eds), *Biological Regulation and Development*, pp 99–125. New York, Plenum Press.

Brown GC, Hafner RP, Brand MD. 1990. A "top-down" approach to the determination of control coefficients in metabolic control theory. *Eur J Biochem* 188:321.

Delgado J, Liao JC. 1992. Determination of flux control coefficients from transient metabolite concentrations. *Biochem J* 282:919.

Epp JK, Huber MLB, Turner JR, et al. 1989. Production of a hybrid macrolide antibiotic in Streptomyces ambofaciens and Streptomyces lividans by introduction of a cloned carbomycin biosynthetic gene from Streptomyces thermotolerans. *Gene* 85:293.

Heinrich R, Rapoport TA. 1974. Linear steady-state treatment of enzymatic chains: General properties, control, and effector strength. *Eur J Biochem* 42:89.

Kascer H, Burns JA. 1973. Control of [enzyme] flux. *Symp Soc Exp Biol* 27:65.

Liao JC, Delgado J. 1993. Advances in metabolic control analysis. *Biotechnol Prog* 9:221.

Malpartida F, Niemi J, Navarrete R, Hopwood DA. 1990. Cloning and expression in a heterologous host of the complete set of genes for biosynthesis of the Streptomyces coelicolor antibiotic undercyl-prodigiosin. *Gene* 93:91.

Mavrovouniotis ML, Stephanopoulos G, Stephanopoulos G. 1990. Computer-aided synthesis of biochemical pathways. *Biotechnol Bioeng* 36:1119.

Mavrovouniotis ML. 1993. Identification of localized and distributed bottlenecks in metabolic pathways. In *Proceedings of the International Conference on Intelligent Systems for Molecular Biology*, Washington.

McAlpine JB, Tuan JS, Brown DP, et al. 1987. New antibiotics from genetically engineered Actinomycetes: I. 2-Norerythromycins, isolation and structural determination. *J Antibiot* 40:1115.

Safer B, Williamson JR. 1973. Mitochondrial-cytosolic interactions in perfused rat heart. *J Biol Chem* 248:2570.

Savinell JM, Palsson BO. 1992. Network analysis of intermediary metabolism using linear optimization: I. Development of mathematical formalism. *J Theor Biol* 154:421.

Small JR, Kacser H. 1993a. Responses of metabolic systems to large changes in enzyme activities and effectors: I. The liner treatment of unbranched chains. *Eur J Biochem* 213:613.

Small JR, Kacser H. 1993b. Responses of metabolic systems to large changes in enzyme activities and effectors: II. The linear treatment of branched pathways and metabolite concentrations. Assessment of the general non-linear case. *Eur J Biochem* 213:625.

Stephanopoulos G, Sinskey AJ. 1993. Metabolic engineering — Methodologies and future prospects. TIBTECH 11:392.

Vallino JJ, Stephanopoulos G. 1989. Flux determination in cellular bioreaction networks: Applications to lysine fermentations. In SK Sikdar et al (eds), *Frontiers in Bioprocessing*, pp 205–219. Boca Raton, Fla, CRC Press.

Varma A, Boesch BW, Palsson BO. 1993. Biochemical production capabilities of *Escherichia coli. Biotechnol Bioeng* 42:59.

Winter RB, Yen K-M, Ensley BD. 1989. Efficient degradation of trichloroethylene by recombinant *Escherichia coli. Biotechnology* 7:282.

Further Information

Good reviews of recent applications of metabolic engineering, as well as advances in metabolic engineering tools can be found in Cameron DC, Tong IT, 1993, Cellular and metabolic engineering: An overview, *Appl Biochem Biotechnol* 38:105; and in Cameron DC, Chaplen FWR, 1997, Developments in metabolic engineering, *Curr Opin Biotechnol* 8/2:175. A pioneering discussion of the emerging field of metabolic engineering can be found in Bailey JE, 1991, Toward a science of metabolic engineering, *Science* 252:1668. A good source of information about recombinant DNA techniques is Gilman M, Watson JD, Witkowski J, et al., 1992, *Recombinant DNA*, 2d ed., San Francisco, WH Freeman.

Index

E

F